History of Analytic Philosophy

Series Editor: **Michael Beaney**, Humboldt University, Berlin and King's College London

Titles include:

History of Analytic Philosophy
Series Standing Order ISBN 978–0–230–55409–2 (hardback)
            978–0–230–55410–8 (paperback)
(*outside North America only*)

You can receive future titles in this series as they are published by placing a standing order. Please contact your bookseller or, in case of difficulty, write to us at the address below with your name and address, the title of the series and one of the ISBNs quoted above.

Customer Services Department, Macmillan Distribution Ltd, Houndmills, Basingstoke, Hampshire RG21 6XS, England

---

Quine and His Place in History

Edited by

Frederique Janssen-Lauret
University of Campinas, Brazil

and

Gary Kemp
University of Glasgow, UK

First published 2016 by
PALGRAVE MACMILLAN

Palgrave Macmillan in the UK is an imprint of Macmillan Publishers Limited, registered in England, company number 785998, of Houndmills, Basingstoke, Hampshire RG21 6XS.

Palgrave Macmillan in the US is a division of St Martin's Press LLC, 175 Fifth Avenue, New York, NY 10010.

Palgrave Macmillan is the global academic imprint of the above companies and has companies and representatives throughout the world.

Palgrave® and Macmillan® are registered trademarks in the United States, the United Kingdom, Europe and other countries.

ISBN: 978–1–137–47250–2

This book is printed on paper suitable for recycling and made from fully managed and sustained forest sources. Logging, pulping and manufacturing processes are expected to conform to the environmental regulations of the country of origin.

A catalogue record for this book is available from the British Library.

Library of Congress Cataloging-in-Publication Data

Quine and his place in history:/[edited by] Frederique Janssen-Lauret, University of Campinas, Brazil, Gary Kemp, University of Glasgow, UK.
    pages cm.—(History of analytic philosophy)
    ISBN 978–1–137–47250–2
    1. Quine, W. V. (Willard Van Orman) I. Janssen – Lauret, Frederique, 1985–editor.

B945.Q54Q46 2015
191—dc23   2015021857

Contents

Part IV   Understanding Quine

Series Editor's Foreword

During the first half of the Twentieth Century, analytic philosophy gradually established itself as the dominant tradition in the English-speaking world, and over the past few decades it has taken firm root in many other parts of the world. There has been increasing debate over just what 'analytic philosophy' means, as the movement has ramified into the complex tradition that we know today, but the influence of the concerns, ideas and methods of early analytic philosophy on contemporary thought is indisputable. All this has led to greater self-consciousness among analytic philosophers about the nature and origins of their tradition, and scholarly interest in its historical development and philosophical foundations has blossomed in recent years, with the result that history of analytic philosophy is now recognized as a major field of philosophy in its own right.

The main aim of the series in which the present book appears, the first series of its kind, is to create a venue for work on the history of analytic philosophy, consolidating the area as a major field of philosophy and promoting further research and debate. The 'history of analytic philosophy' is understood broadly as covering the period from the last three decades of the Nineteenth Century to the start of the Twenty-first Century, beginning with the work of Frege, Russell, Moore and Wittgenstein, who are generally regarded as its main founders, and the influences upon them, and going right up to the most recent developments. In allowing the 'history' to extend to the present, the aim is to encourage engagement with contemporary debates in philosophy, for example, in showing how the concerns of early analytic philosophy relate to current concerns. In focusing on analytic philosophy, the aim is not to exclude comparisons with other – earlier or contemporary – traditions, or consideration of figures or themes that some might regard as marginal to the analytic tradition but which also throw light on analytic philosophy. Indeed, a further aim of the series is to deepen our understanding of the broader context in which analytic philosophy developed, by looking, for example, at the roots of analytic philosophy in neo-Kantianism or British idealism, or the connections between analytic philosophy and phenomenology, or discussing the work of philosophers who were important in the development of analytic philosophy but who are now often forgotten.

Willard van Orman Quine (1908–2000) was one of the leading figures in the second generation of analytic philosophers. Born in Akron, Ohio, he studied mathematics at Oberlin College before doing his PhD under the supervision of A.N. Whitehead at Harvard on Whitehead and Russell's *Principia Mathematica* 1910–1913. He spent the academic year of 1932–33 in Europe, taking part in meetings of the Vienna Circle and visiting Rudolf Carnap (who was in Prague at the time). He returned to Harvard in 1933 as Junior Fellow and remained there for the rest of his life, becoming Professor in 1948 and Edgar Pierce Professor of Philosophy in 1956 until his retirement in 1978. His most important works include *Mathematical Logic* (1940), *From a Logical Point of View* (1953), which contains two of his most famous papers, 'Two Dogmas of Empiricism' and 'New Foundations for Mathematical Logic', *Word and Object* (1960), *The Ways of Paradox and Other Essays* (1966, 1976), *Ontological Relativity and Other Essays* (1969), *Theories and Things* (1981), and *Pursuit of Truth* (1990).

'Two Dogmas of Empiricism', first given as a paper in 1950, heralded the critique of logical empiricism that was such a central feature of analytic philosophy in the 1950s and 1960s. The two dogmas that Quine attacked were the analytic–synthetic distinction and reductionism, and Carnap's views were certainly one of the targets. Quine came to develop a form of naturalism, in which philosophy was seen as continuous with natural science, and to take seriously ontological questions, which led to the return to metaphysics after its repudiation by both logical positivism and ordinary language philosophy. All of these themes are addressed and explored in the present collection, edited by Frederique Janssen-Lauret and Gary Kemp, especially in Part IV.

As the editors note in their Introduction, it is now time for detailed historical study of the development of analytic philosophy in the second half of the Twentieth Century. In the case of Quine, this involves not just investigation of the intricacies of his engagement with logical empiricism but also of the influence upon him of pragmatism, which has been a powerful tradition – and arguably, the dominant tradition – in American philosophy throughout the Twentieth Century. There are two papers on Quine's connection to pragmatism in Part III, and an interesting account of Quine's contact with the Unity of Science Movement, integral to the logical empiricist tradition, in Part II. We are also delighted to include some previously unpublished papers by Quine, with accompanying commentary, in Part I. With this volume, history of

Quinean philosophy can be seen not only to have come of age but also to have taken its rightful place in history of analytic philosophy, with which it is undoubtedly continuous.

Michael Beaney
June 25, 2015

Acknowledgments

This volume of papers grew out of a joint Glasgow–Campinas conference on Quine held in Glasgow in December 2014. For this conference we gratefully acknowledge the support of the Scots Philosophical Association, the Mind Association, and the Philosophy departments at the Universities of Glasgow and Campinas, as well as the Centre for Logic and Epistemology at the University of Campinas. We're also grateful to Jane Heal, Fraser MacBride, and Alan Weir for presenting papers at the conference and providing further support and advice on the project, to Douglas Quine for presenting an enlightening illustrated account of his father's life and works there, and to Berta Grimau, Nathan Kirkwood, and Finlay McCardel for further help at the conference.

We owe a great debt to Rolfe Leary and Gary Ebbs for allowing us to include in this volume the previously unpublished W.V. Quine papers in Part I. We are also most grateful to Douglas Quine for his many helpful suggestions, for his design of the front cover, and for granting us permission to reprint 'Levels of Abstraction', 'Preestablished Harmony', and 'Response to Gary Ebbs'.

Thanks are also due to Michael Beaney, the editor of this series, and to Esme Chapman and Brendan George, Philosophy editors at Palgrave Macmillan, for their advice and support in making this volume a reality. Many thanks to Berta Grimau for compiling the index.

The editorial work on this volume was partly supported by a Capes Postdoctoral Research Fellowship Grant.

Notes on Contributors

Yemima Ben-Menahem is Professor of Philosophy at the Hebrew University of Jerusalem, working in particular in the philosophy of science. She is the author of *Conventionalism* (2006), editor of *Hilary Putnam* (2005) and co-editor of *Probability in Physics* (2011).

Gary Ebbs is Professor and Chair of Philosophy at Indiana University, Bloomington. He is the author of *Rule-Following and Realism* (1997), *Truth and Words* (2009), and (with Anthony Brueckner) *Debating Self-Knowledge* (2012), as well as a number of articles on topics in the philosophy of language and the history of analytic philosophy.

Peter Hylton is Professor of Philosophy and UIC Distinguished Professor at the University of Illinois, Chicago. He has published extensively, chiefly on the history of analytic philosophy.

Frederique Janssen-Lauret is a postdoctoral research fellow at the University of Campinas, Brazil, working on philosophy of logic and the history of analytic philosophy. She has published papers on Quine and meta-ontology in *Synthèse* and *The Monist*.

Gary Kemp is a senior lecturer at the University of Glasgow. He is the author of several papers on Quine (and on Davidson, Frege, and Russell), and the books *Quine versus Davidson: Truth, Reference and Meaning* (2012) and *Quine: A Guide for the Perplexed* (2006).

Rolfe A. Leary worked 35 years as a research scientist with the USDA Forest Service Research developing mathematical models of the dynamics of Northern USA forests. He began collaborating with Edward Haskell in 1970, and wrote a book about his application of Haskell's ideas to his work in 1985; raleary@comcast.net.

Ann Lodge is a clinical child psychologist who has conducted research on attachment and intervention in infancy, as well as behavioral and electrophysiological aspects of early development. She has served on the faculties of Eastern Virginia Medical School, George Mason University, Old Dominion University and the University of California, San Francisco; analog123@aol.com.

Andrew Lugg, a professor emeritus, who lives and works in Montreal. The author of *Wittgenstein's Investigations 1–133*, he is presently writing a book on Wittgenstein and color.

Douglas B. Quine was an associate professional scientist at the Illinois Natural History Survey/University of Illinois in 1988 studying bird migration when his passion for improving postal automation systems led to a career in business. He retired as research fellow at Pitney Bowes Advanced Concepts and Technology with 49 US patents and currently consults while managing the W.V. Quine literary estate; drquine@gmail.com.

W.V. Quine arrived at Harvard University in 1930, earned his PhD in two years, and died in 2000 as the Edgar Pierce Professor of Philosophy Emeritus. His prodigious eloquent literary output in mathematical logic, set theory, epistemology, and the philosophy of language was recognized worldwide and inspired this Glasgow Conference.

Robert Sinclair is Associate Professor of Philosophy in the Faculty of International Liberal Arts, Soka University, Tokyo. His work examines themes from the American pragmatist tradition focusing especially on the philosophies of John Dewey and W.V. Quine.

Introduction: Quine and His Place in History

Frederique Janssen-Lauret and Gary Kemp

I

A central aim of the historical study of philosophy is to gain a certain type of intellectual self-consciousness. Retracing the paths of our forbears, we see decisions being made, sometimes tacitly or implicitly; we see the routes not taken and often the reasons why; confusions avoided or fallen into and insights won or lost; we gain a sense of things we now take for granted as optional. We learn more about who we are.

This point holds all the more for the historical study of analytic philosophy by analytic philosophers. Of course analytic philosophers of a historical frame of mind have long displayed extensive interest in Frege, Russell, Moore, Carnap, and the early Wittgenstein. They've become increasingly aware of and interested in the history of their discipline, turning their thoughts to key philosophers of various established branches of analytic traditions, including logicism, logical positivism, Wittgensteinianism, and pragmatism; those views became less the order of the day and more the products of their particular time and place, and therefore proper objects of historical study.

But the historical study of analytic philosophy was until recently confined to the early stages of its development. Now that the Twentieth Century has given way to the Twenty-first, the field is broadening to include not just the earliest beginnings of analytic philosophy, but the mid-Twentieth Century. And one of the pivotal figures of this epoch is W.V. Quine (1908–2000). Many analytic philosophers now at work came of age only after the publication of his final two books in the 1990s; their teachers in turn came of age when his celebrated early works were already receding into the past. And the point made in the opening paragraph looms especially large when it comes to Quine. For all that Quine's output is voluminous, Quine's work is above all systematic; and the systematic nature of his work is largely lost on the student struggling to cope with individual works such as 'Two Dogmas of Empiricism', 'Quantifiers and Propositional Attitudes', or the second chapter of *Word and Object*. It's too big, and too alien. Despite Quine's being a seminal figure in analytic philosophy, much of his work stands opposed to the framework – possibly merely tacit – in which the analytic philosopher is trained and works. There is a real danger of the student's thinking of herself as a follower of Quine without understanding what it means to say so. More historical awareness of Quine is urgently needed.

Not that this is a thoroughgoing exegetical and historical study of Quine in all philosophical aspects. Quine's famous intellectual relationship with Carnap, which began in earnest with Quine's 1933 visit to Carnap in Prague, has already been examined in detail, notably by Richard Creath in his *Dear Carnap, Dear Van* (1991). Nor have we touched on Quine's career as a logician and set theorist; but of course that subject by its nature is much less susceptible to the obscuring mists of history (the set theory of Quine's 'New Foundations for Mathematical Logic' remains a live research topic; see Randall Holmes' *New Foundations Home Page*, http://math.boisestate.edu/~holmes/ holmes/nf.html). More generally we take for granted the reader's knowledge of the basics of Quine's career (for those not satisfying that condition, we recommend Quine's compact Intellectual Biography in the Schilpp volume on Quine in the *Library of Living Philosophers* (1982); for those wanting more, his book-length autobiography – *The Time of My Life* (1985) – expands on the Intellectual Biography); and we take for granted the reader's grasp of the very basics of Quine's philosophical system. Our primary aim here is to fill in some major gaps in the historical narrative, scholarship and exegesis of Quine. This volume of papers on Quine and his historical context brings together notable Quine scholars from around the world to provide their different perspectives upon the development of Quine's philosophy, the philosophers and scientists who influenced him, and some of the ways in which historical investigation can shed light upon the details of his accounts of language, knowledge, and metaphysics (or his attitude towards metaphysics). It also provides certain papers with a fine-grained exegetical purpose, which it is hoped will not only answer some important and lingering interpretational questions, but serve the above aim of our seeing more clearly our historical position, of furthering our intellectual self-consciousness.

II

We feel very fortunate to be able to present to the world, in Part I of this volume, three previously unpublished short papers by W.V. Quine. Little did we suspect, when we sent out a call for papers, that the eventual book would feature not just one, but three posthumous pieces from the hero of our tale. The first paper, 'Levels of Abstraction', was generously provided by Rolfe Leary, keeper of the Nachlass of Ed Haskell. Quine was a formative influence on the Unity of Science movement and a close friend of Haskell, who was himself the founder of the Council for Unified Research and Education, a defender of some of the key principles of pragmatism, and a formidable proponent of his own distinctive form of scientific realism (others active in the movement include Philip Frank, Otto Neurath, Charles Morris, and, if somewhat reluctantly, Rudolf Carnap). Haskell's relation to Quine is discussed in this volume by Ann Lodge, Rolfe Leary, and Douglas Quine. Haskell had not only been one of the instigators of the Unity of Science movement, but he was also Quine's housemate while they were undergraduates at Oberlin College. Haskell went on to postgraduate study at the University of Chicago, where Leary (in conversation) hypothesizes he came across Neurath, Carnap, and Morris doing research into the Unity of Science. He organized a symposium on the theme in 1948 at the American Association for the Advancement of Science. From this event sprang the formation of a loose-knit group of sympathizers, drawn from across several disciplines, meeting up at irregular intervals over the years under the banner of CURE (Council for Unified Research and Education). In 1972, Haskell, having made contact with the Unification Church (the 'Moonies'), used their financial support to host the First International Congress on Unified Science in grand style at the Waldorf Astoria in New York City. Quine, by this point rather skeptical of Haskell's Unified Science project, as well as of organized religion, reluctantly agreed to give a paper on abstraction. In the audience was a mathematically inclined research forester and supporter of unified science, Rolfe Leary. He took his copy of Quine's handout home with him, and stored it in a filing cabinet in the house he shared with the psychologist, and fellow member of the Unity of Science movement, Barbara Buckett Leary. For the next 42 years, it was assumed that no copies of the paper had survived at all, until Douglas Quine found out about the existence of Leary's copy. Douglas Quine has transcribed and edited the original typescript, not typed by W.V. Quine himself, which contained several inserted errors.

Two further papers, dating from the mid-1990s, were kindly bestowed upon us by Gary Ebbs. The first is a short draft paper responding to Ebbs' review of Quine's *Pursuit of Truth*, the second a revision of it which shows an intriguing glimpse into the usually covert influence upon Quine of Burton Dreben. These two papers were typed by W.V. Quine on his trusty old typewriter which appears in our cover image, many of whose standard-issue keys he had replaced with logical symbols. Since this means the originals are of historical interest, scans of them appear in our Appendix. The main text of the book contains versions of these two papers edited and transcribed by Gary Ebbs. These letters and manuscripts were reprinted with the permission of Dr. Douglas Quine, W.V. Quine Literary Estate.

Part II provides a historically interesting glimpse into Quine's complex relationship with Haskell and the Unity of Science movement. This paper's authors saw events unfold in real time. Ann Lodge, a psychologist, was married to Haskell for several years, and was also the daughter of G.T. Lodge, also a psychologist and central member of the Unity of Science movement. Rolfe Leary, the literary executor and regular correspondent of Harold Cassidy and Ed Haskell, is the keeper of these two men's literary estates and is currently in the process of editing a volume of their collected works, begun by Haskell and Cassidy but also incorporating works by Quine and other collaborators.

Lodge, Leary, and D. Quine draw upon the extensive correspondence between W.V. Quine and Haskell, as well as correspondence with other members of the movement such as G.T. Lodge, and the brothers Fred and Harold Cassidy, to paint a picture of Quine's influence upon that movement. Although the movement had its roots in a meeting of minds between these men while they were students at Oberlin College, overall Quine's contributions consisted mostly of tempering Haskell's exuberant optimism. Haskell had high hopes, not just for finding a set of classificatory principles applicable in equal measure to social and natural science, but also for deriving normative insights from such principles to cure the world's ills. Quine grew increasingly skeptical of Haskell's efforts, and subsequently frustrated with them. Still he persisted in reading his old friend's work and offering suggestions, urging him towards a better informed conception of mathematical rigor and clearer distinctions between unification at the level of explanation versus description, and thereby perhaps exerting a sobering influence.

Many of the papers in Parts III and IV derive novel insights from negotiating intersections between Quine and other significant thinkers of the late Nineteenth and early to mid-Twentieth Century – to some familiar giants of analytic philosophy (Wittgenstein, Russell, James, Peirce), and to some comparatively under-researched, like C.I. Lewis and Ruth Barcan Marcus. Another theme that is shared between several of the contributions to this volume is the historical context and development of Quine's naturalism, considered from different angles: its connection to pragmatism, potential challenges to or from scientific realism, and Quine's replies to alternative versions of naturalism such as those offered by the Unity of Science movement or classic nominalism. Still we've separated Part III from Part IV according to genre: Part III is more purely historical; it contains papers on Quine's relationship to his pragmatist forebears and on the younger Quine in dialogue with his pragmatist and Unity of Science contemporaries. Part IV is more exegetical and critical; it concerns some especially difficult or insufficiently noted aspects of Quine, though still frequently by comparison to other historical figures.

Ben-Menahem considers Quine's pragmatist epistemic holism in connection with the views of James. She argues that similarities between the two have been overlooked owing to a widespread misinterpretation of James as holding that there is nothing to truth and rationality except usefulness, and that the differences between them are largely due to the different kinds of positivism each was responding to. She aims to locate Quine more squarely in the pragmatist tradition dating back to James by elucidating affinities between Quine's and James's views on metaphysics, skepticism, and the social dimension of knowledge.

Sinclair's paper traces Quine's pragmatism to a previously unremarked source: the influence of Quine's postgraduate supervisor C.I. Lewis. Focusing on the pragmatist conception of the a priori which is a key component of Lewis's work, Sinclair examines Quine's unpublished student work for signs that the early Quine employed Lewis's view, attempting to modify it to suit his own needs in a way that foreshadows developments in the mature Quine.

Hylton discusses the seldom observed split in Quine's philosophy of language between ontology and regimentation, on the one hand, and the understanding of language on the other. The split is revealingly contrasted with the philosophy of language of Russell, for whom the notion of acquaintance provides the meeting point: what is required for the understanding of a sentence is precisely acquaintance with those entities which must exist for the sentence to be meaningful. For Quine, these are different subjects: The understanding of a sentence is just the having of certain linguistic dispositions, and does not require awareness of reference or ontology. The latter are scientific or technical subjects, involving regimentation, into mere first-order predicate calculus, of scientific theory.

Ebbs offers an alternative reading of Quine's famous claim in 'Two Dogmas of Empiricism' that no statement is immune to revision. He notes that fans and detractors of Quine alike generally interpret this as meaning, as he puts it, that 'for every statement S that we now accept, there is a possible rational change in beliefs that would lead one to reject S'. Ebbs argues that this standard interpretation fails to take account of Quine's views on translation, which problematize the idea of homophonic translation on which the standard interpretation relies, and that it is at odds with the context in which the claim is made, in which there is no reference to homophonic translation or belief revision. He proposes, instead, that Quine's aim in section 6 of 'Two Dogmas of Empiricism' where he makes his claim that no statement is immune to revision, is to propose a naturalistic revision of the notion of empirical confirmation. The claim itself, in its proper context, is linked to Quine's efforts to make clear that empirical confirmation as he conceives of it, as opposed to the traditional notion, is not conducive to dividing statements into the analytic and the synthetic. So Ebbs puts forward an improved reading of Quine's claim: 'No statement we now accept is guaranteed to be part of every scientific theory that we will later come to accept'.

Janssen-Lauret explores the diametrically opposed nominalistic naturalisms of Quine on the one hand and Ruth Barcan Marcus on the other. While both favor an ontology composed entirely or primarily of concrete physical particulars, their epistemological motivations for this choice and their respective meta-ontologies differ radically. For Quine, ontological commitments must always be analogous to positing in science: existential assumptions result from solutions to questions about the best overall descriptions that fit our observational patterns. Barcan Marcus, by contrast, thinks of physical particulars as encounterable, and nameable, directly via knowledge by acquaintance. The paper examines their resulting differences in their interpretations of quantification and identity.

Kemp considers the apparent tension between two commitments in Quine: his Realism, and the Underdetermination of Theory. On the face of it, it seems that one cannot hold that our wholesale account of nature could in principle be exchanged for another, wholly different account of nature, without impugning one's claim that our actual account provides us with knowledge of nature, nature as it really is. As so often is the case with Quine, the Quinean resolution involves his naturalism, and in particular his naturalistic account of language. But it is a delicate balance; to maintain it requires a careful coming to terms with the concepts of transcendental metaphysics, of words such as 'reality', 'the world', 'existence', and the like.

Lugg considers the influence of Quine's scientism in his attitude towards the 'abyss of the transcendental', attempting to rescue what we can from the chasm, by contrasting this attitude with Wittgenstein's complementary but opposing attitude of diving straight into the abyss and exploring the transcendental territory. Lugg aims to shed light upon the deep methodological differences between Quine and Wittgenstein by exploring their different attitudes here, and argues that Quinean and Wittgensteinian approaches are not incompatible, but can each in their way guide other thinkers who are skeptical of the transcendental.

Part I

Previously Unpublished Papers by W.V. Quine

1
Introduction to 'Levels of Abstraction'

Douglas B. Quine

Forty-two years ago, as a biology student and news reporter for the Princeton University radio station WPRB, I sat in the audience at the Waldorf Astoria hotel in New York City and heard my father give a talk entitled 'Levels of Abstraction'. Since then, I knew of no copy of the text of that talk in any journal, book, university, or family archive. Even the title of the talk was forgotten until two months ago when Dr. Rolfe Leary, a co-author of our companion paper at this conference, casually mentioned that he had retained a copy of Quine's paper from the New York conference. He provided the typewritten preprint to me which I transcribed last month in Antarctica – the one continent that my father never visited. Philosophy and mathematics often lead the way for computer science and I believe this paper takes on a new level of relevance in an era of computer programming and big data. It is with great pleasure that I present the unpublished 'Levels of Abstraction' today and provide it finally for publication in the proceedings of this conference.

2
Levels of Abstraction (1972)
W.V. Quine

Levels Of Abstraction[1]

Some terms are more abstract than others. Some terms are not more abstract than any others, and they constitute the zero level of abstraction. Some terms are more abstract than those of zero level, but not more abstract than any others, and they constitute the first level of abstraction. Some are more abstract than those of level one and zero, but not more abstract than any further ones; and they constitute the second level of abstraction. And so on up. I seem thus to have defined the levels of abstraction, but it is not much of a definition, for it assumes that we know what it means for one term to be more abstract than another. This I shall not define, but I shall point out some confusions over it.

Is the word 'mammal' more abstract than 'rodent'? Is 'rodent' more abstract than 'mouse'? Is abstractness thus merely a question of inclusiveness? Surely not. Surely 'apple' is not more abstract than 'winesap', nor 'sugar' more abstract than 'levulose'. Inclusiveness is one thing, abstractness another.

Sometimes what is conjured up by talk of abstraction is rather the hierarchy of naming. At the bottom there are things; next above them there are names of things; next there are names of those names; and so on up. Lewis Carroll touched on this.

"The name of the song is called 'Haddocks' Eyes'." [2] [3]
"Oh, that's the name of the song, is it?" Alice said,
trying to feel interested.
"No, you don't understand," the Knight said, looking
a little vexed. "That's what the name is called.
The name really is 'The Aged Aged Man'."
"Then I ought to have said 'That's what the song
is called'?" Alice corrected herself.
"No, you oughtn't: that's quite another thing!
The song is called 'Ways and Means': but that's
only what it's called, you know!"
"Well, what is the song then?" said Alice, who
was by this time completely bewildered.
"I was coming to that," the Knight said.
"The song really is 'A-sitting On a Gate'."

A more venerable example is the tetragrammaton.
This fourteen-letter word 'tetragrammaton' was the
name of a four-letter word, and its utility came of
the circumstance that the four-letter word was taboo.
The four-letter word[4] was spelled yodh, he, vav[5], he,
reading from right to left in the Hebrew fashion, and
it was pronounced Yahweh (soll mir nicht schuldigen).
Here then we have three levels, if for the sake of the
example you will grant me the existence of God. At the
bottom level we have the Deity Himself. (I do hope I
can get through this part without being struck down by
a bolt of lightning[6].) At the next level we have his
four-letter name, 'Yahweh'. At the third level we have
its fourteen-letter name, 'tetragrammaton'. And we can
ascend to further levels, uninterestingly, by applying
quotation marks and more quotation marks.
  This form of hierarchy – the thing, the name, the
name of the name – is generally sterile after the
first two levels. The initial distinction[7], between
things and their names, is important. But when we talk
about the names we ordinarily form names of names
by simple quotation, and rise no higher. The case of
'tetragrammaton' was a rare case, and due only to a
strange taboo.

This hierarchy becomes somewhat richer if instead of
limiting ourselves to <u>designation</u>, by singular names,
we look also to <u>denotation</u> by general terms, that is,
common nouns and adjectives. Three levels then stand
vividly forth. At the zero level as always there are
the things, God bless them. At the next level there are
now not only the names of things, names like 'Boston'
and 'Washington Monument' and 'Bernard J. Ortcutt'[8],
but also there are general terms like 'rat', 'rodent',
'mammal', 'city', 'monument', etc. that apply to
things. At the level above these there are not only
names of names, like 'tetragrammaton', but also general
terms that apply to names and to general terms. There
are such general terms as the word 'noun', which
applies to the words 'rat' and 'city' and all the rest;
also the word 'trochee', which applies to the words
'Boston', 'rodent', 'city', and 'mammal'; also the word
'dactyl', which applies to the word 'monument'.

This hierarchy thus rejoices in three lively levels,
and lapses into dullness above these. It turns out,
moreover, that these levels must not be confused, on
pain of paradox. The paradox was propounded by Kurt
Grelling 65 years ago. It hinges on the general terms
'autological'[9] and 'heterological'. These are terms
of level two, if we count things as of level zero.
These terms are at the level of the words 'noun' and
'trochee' and 'dactyl': words of level two, applicable
to words[10], of level one. Here is what they mean:
a word is <u>autological</u> if it is truly applicable to
itself. Thus the word 'short' is autological, being a
short word. The word 'English' is autological, being
an English word. The word 'word' is autological, being
a word. The word 'trochee' is autological, being a
trochee. Other words are called <u>heterological</u>; thus
the words 'long' and 'German' are heterological, not
being long or German. And now here comes Grelling's
paradox: is the word 'heterological' autological or
heterological? If the word 'heterological' is itself
heterological, and thus true of itself that makes it
autological; and vice versa. By respecting the levels
of our hierarchy, however, we dodge the paradox. The

second-level word 'heterological' is applicable only
to first-level words, and we confuse levels when we
ask whether it is itself heterological.

So we see that these linguistic levels of denotation
do matter; the words that apply to words occupy
a significantly higher level than the words that
apply to things. Still this hierarchy also, however
important, fails to capture what one wants to call
the levels of abstraction. Abstractness is not a
matter of inclusiveness, we saw, and it is also not
a matter of loftiness in the hierarchy of names of
names of names. It may be better identified, surely,
with yet a third kind of thing: it is a question of
classes and classes of classes, or of properties and
properties of properties. Thus consider again the mice
and the rodents. The mice constitute a zoological
family; the rodents a zoological order. Each mouse,
indeed each rodent, is a thing of abstraction level
zero. The mouse family, taken as a class or property,
belongs to the next level of abstraction; and so
does the order Rodentia. The order Rodentia is more
inclusive than the mouse family, but they are both at
the same level of abstraction, namely level one. And
then at level two we have the class of all zoological
families, or the property of being a family; likewise
the class of all orders; the class of all species; and
also the union of all these classes, hence the class
of all taxa, as taxonomists call their taxonomical
categories. Thus each mouse, each chipmunk, each
rodent, each individual, belongs to the zero level.
The mouse family taken as a class, or property,
belongs to level one; and so does the order Rodentia,
and so does the class Mammalia, and the phylum
Chordata. Each of these, some more inclusive and some
less, stands at the first level of abstraction. Then
at the second level we have classes of classes, or
properties of properties. One of them is the class of
all families, one is the class of all species, one is
the class of all orders, one is the class of all taxa
of all sorts. At the third level of abstraction we
have classes of such classes of classes; and so on up.

These levels of abstraction are what Russell called
types. They are levels of abstraction in a serious
sense. Significantly enough, this hierarchy also goes
rather dim after a few levels. This is a commentary
on our own limitations in relation to levels of
abstraction. Mathematicians concerned with abstract set
theory blithely encompass an infinite hierarchy of such
levels, and even emerge into transfinite levels. Well,
they are intelligent men, but let us not overestimate
them. They gain their swift ascents by formulating
the general principle of ascent and not fretting over
specific applications. Conversely our own adherence to
a few bottom levels of this hierarchy can be accounted
for in terms of utility and diminishing returns.

The two hierarchies that I have last described are
sometimes conflated, and we can see why. One is a
hierarchy of words and phrases, in which those of
higher level are true of those of lower level. The
other hierarchy, the types, is a hierarchy rather of
classes or properties in which those of higher level
are classes or properties of those of lower level.
Practical men of an amiably nominalist bent are apt
to view the classes or properties as mere words or
phrases, and thus to identify the hierarchy of classes
or properties with the hierarchy of words and phrases.
However, there are differences. There are logical
differences of a technical kind that I shall not
pause over, and there are common-sense differences.
It would be awkward to identify the mouse family with
the word 'mouse', or the order Rodentia with the word
'rodent', even apart from technical troubles. We want
to say of the word 'mouse' that it is short and of the
word 'rodent' that it is a trochee; we want to say
quite other things of the mouse family or the order of
Rodentia – e.g., that it is numerous.

The nominalist urge to reduce abstract objects to
mere abstract words is both amiable and understandable.
For how, one may ask, can people learn to talk about
abstract objects – classes, properties – when only
concrete objects are present to the senses? This is a
good question and I think it admits of a good answer,

though not a brief one. We can reconstruct plausible steps whereby people can have learned to talk not only of observable concrete objects but also of abstract ones. Some of the steps proceed by conspicuous analogy and unconscious extrapolation. Some of them depend on confusions. Confusion of sign and object. Confusion, also, of concrete general term with abstract singular. A priori the steps are not justifiable. A posteriori they gain pragmatic justification: our scientific conceptual scheme is a going concern, and no rigorous way is known of ridding it of sets and numbers and functions and other abstract objects, in form of mere words. The flat-footed way, simply saying that the sets and numbers and functions are mere words, runs into technical snags - I repeat - that are not to be analyzed here. But the point that I would bring out is that notions that were bred by confusion can admit still of pragmatic justification; happy accidents are happy even though accidental. Conceptual schemes, like species, evolve by natural selection on the strength of their survival value; and the inception of a conceptual scheme need be no more rational than a genetic mutation. Rationality deserves encouragement, certainly, but let us not give up our happy accidents.

The hierarchy that we last arrived at, that of types, is decidedly a hierarchy of levels of abstractions. At the bottom there are individuals. Next above there are classes or properties of individuals; also, I should add, relations of individuals. Next there are classes or properties and relations of such classes or properties, and relations; and so on up.

But there remains still another dimension of abstraction, which I have not yet touched upon, that may be psychologically much more significant still. It has to do with cyclic principles of generation, and cyclic principles of generation of cyclic principles of generation, and so on up.

Thus consider how we count. We can go on and on. Thanks to a cyclic principle. Our primitive forebears counted mangoes on their fingers and thus counted to ten mangoes. Our own trick consists in putting the

tens in place of the mangoes the next time around, and
thus making a hundred: ten tens. Once we have the trick
of putting the tens for the mangoes, we don't stop
at that; we put our newly acquired hundreds for them,
and thus make a thousand: ten hundreds. Here, surely,
is abstraction at its best and its most significant:
application of our generation to itself. The operation
of the ten-count was suited at first to the mangoes,
but then we apply it to the ten-count itself, getting
the hundred-count; and then to the hundred-count[11], and
so on up.

Our primitive forebears missed this bet, and so got
little beyond their ten fingers. They managed eleven
and twelve, calling them 'one left over' and 'two
left over'; and [that][12] is the etymology of our words
'eleven' and 'twelve', or 'elf' and 'zwoelf'. I speak
of the wild Germanic tribes. Other primitive peoples
have been known to count 'eight', 'nine', 'ten',
'jack', 'queen', 'king'. But the Romans were on to the
cyclic trick, with 'undecim', 'duodecim', 'tredecim',
and so on. Meanwhile some primitive cousins of our
primitive forebears forged ahead well beyond eleven
and twelve by counting fingers and toes, and calling
it a 'score'; but without recycling. We did better
by learning to recycle our tens before resorting in
desperation to our toes. Our decimal system is more
sophisticated than vigiatesimal; it uses fewer stages
to get into orbit. And it could be argued that the
system of numeration based on 2, the lowest possible
base, is the most sophisticated of all. It flips into
cycles at every turn; cycles are everything. It is
simplest, theoretically, for it proceeds by binary
choices; the choice is always this or that rather than
one in ten. Hence the importance of base-2 numeration
for the theory of computers.

Mathematicians who work in abstract set theory have
been cycling into the empyrean. They not only apply
an operation to itself, they apply it to the very
operation of applying it to itself, thus attaining a
higher level of abstraction in quite an exalted sense.
And then bring all this to bear upon itself, and so on

up. Each successive level of this hierarchy is utterly vertiginous. This is how the set theorists get their transfinite ordinal numbers.

I again hasten to say in behalf of common sense that these set-theoretic flights of virtuosity[13] are studies in flight for flight's sake with no fretting over specific applications. But I think still, putting these excesses aside, that the cyclic feature is[14] the key to levels of abstraction in the best sense, the sense that is most relevant to intellectual levels and intelligence quotients. Think how central the cyclic feature is to any ordinary mathematics, also quite apart from the flights of set theory. Mathematical induction is the vital principle of number theory; it was recognized as such by Poincaré and centuries earlier, by Fermat[15], and must be so recognized by us all. And it is vividly cycled, in unit cycles; we prove a law by proving that it holds of 0 and that if it holds of any number, it holds of the next. This principle of mathematical induction is the fundamental principle of number theory. For that matter, the very use of parentheses and variables – so characteristic of all mathematics – serves largely to pave the way for cyclic reasoning. The parentheses enable us to gather any highly complex expression together as a single unit and substitute it for a simple variable in a given formula; and the complex expression which is thus substituted in the given formula may contain a replica of that given formula itself, or consequences of it. This is where mathematics gets its power. This is what generates its escape velocity. Mathematics is the medium <u>par excellence</u> of cyclic reasoning. Cyclic reasoning is reasoning at its most powerfully abstract. And mathematics is science at its most abstract. Mathematics is not necessarily the most fruitful part of science, and the most abstract part of mathematics is perhaps not the most fruitful part of mathematics. But I expect there is a correlation between the intellectual level of a culture, in some significant sense of intellectual level, and the abstractness of its mathematics.

Notes

1. An unpublished paper presented at the *First International Conference on Unified Science,* NY, November 23–26, 1972. A typewritten preprint of "Levels of Abstraction" was provided by Dr. Rolfe Leary during the writing of the companion paper [Lodge, Leary, Quine] in this volume which describes the conference. Previously no copy of the paper was known to the Quine archives.
2. The preprint manuscript was professionally typed on a new typewriter – not on Quine's 1927 Remington. The single character typographical errors referenced in the editor's footnotes below may have been introduced by the conference typist in 1972 and not detected by Quine at the time.
3. Editorial correction to Haddocks' (plural possessive) to match Lewis Carroll *Through the Looking Glass*
4. יהוה (enabled by computer technology 42 years later)
5. Editorial correction of "van" (perhaps a Freudian slip) to the correct Hebrew letter "vav"
6. Editorial correction of "lightening"
7. Editorial correction of "distraction"
8. Editorial correction of "Orteutt" to the familiar "Ortcutt" in Quine's writings (e.g. 1989, 1992, 1994, and 1995)
9. Editorial correction of "antological" which appears throughout the preprint (see note 2)
10. Editorial correction of "word"
11. Editorial expansion: "and then to the hundred-count, getting   the thousand-count"
12. Editorial addition
13. Editorial correction of "virtuousity"
14. Editorial change from "be"
15. Editorial change from "Fermet"

3
Introduction to 'Preestablished Harmony' and 'Response to Gary Ebbs'

Gary Ebbs

In July 1995, I was surprised and delighted to receive by US mail from W.V. Quine a draft of a short paper titled 'Preestablished Harmony' that responds to Ebbs (1994), my review of Quine (1992), the revised edition of his book *Pursuit of Truth*. A few days later Quine sent me another short paper titled 'Response to Gary Ebbs' and a brief cover letter in which he wrote:

> Here is a revision of the four pages I sent you last week. It was sparked by discussions with Burt Dreben. I mean to submit it or a further version to the Phil Review.

The Philosophical Review, which does not publish replies to its reviews, did not publish Quine's paper. One year later, however, *The Journal of Philosophy* published a substantially revised version of the central points of the paper (now framed without mention of my review) as 'Preestablished Harmony', the first section of his paper, 'Progress on Two Fronts' (Quine, 1996).

Soon after receiving 'Response to Gary Ebbs', I sent a letter to Quine explaining my remaining puzzlement about his evolving views; he then sent me a detailed reply. Together with 'Preestablished Harmony' and 'Response to Gary Ebbs', Quine's reply to my letter sheds light on his evolving views of the intersubjectivity of observation sentences during the period between the publication of *Pursuit of Truth* and his 1996 paper 'Progress on Two Fronts'. The differences between 'Preestablished Harmony' and Quine's revision of it, 'Response to Gary Ebbs', which, according to Quine, was 'sparked by discussions with Burt Dreben', also afford a rare, though indirect, glimpse of Quine's working relationship with Dreben, whose influence on Quine's work is in most cases more difficult to discern. With the thought that these previously unpublished documents may be of use to Quine scholars and others who are interested in the issues they concern, in this introduction I sketch their contexts and contents in chronological order, without critical commentary. Typeset and lightly edited versions of 'Preestablished Harmony' and 'Response to Gary Ebbs' directly follow this introduction. Facsimiles of Quine's original type-written copies of 'Preestablished Harmony' and 'Response to Gary Ebbs' (copies he sent to me in July 1995) appear in the Appendix of this volume.

1   Excerpts from my review of *Pursuit of Truth*

I begin with some excerpts from my review of *Pursuit of Truth*, which prompted the exchange. In the review I argue that in *Pursuit of Truth*, Quine radically changes his view of the relationship between stimulus meaning and intersubjectivity. I summarize the change as follows:

> One of the distinctive features of Quine's naturalized epistemology was that the intersubjectivity of observation sentences was mirrored by intersubjective similarity of stimulus meanings. This provided a link between intersubjectivity and Quine's naturalistic descriptions of intrasubjective associations between sentences and sensory stimulations. It also figured centrally in Quine's account of what is objective in translation. In *Word and Object* Quine held that translation of an observation sentence S is a matter of finding a sentence of the translator's language that has the same stimulus meaning as S. But in *Pursuit of Truth*, Quine observes that by his definition the patterns of sensory stimulation that each subject receives depend on idiosyncratic anatomical details about her nerve endings. He argues that any criterion for intersubjective similarity of stimulations would rest on "anatomical minutiae" that "ought not to matter here." [*Pursuit of Truth*, p. 40.] Instead of trying to save the idea of intersubjective stimulus meanings, Quine concludes that "we can simply do without it." [*Pursuit of Truth*, p. 42.] He now holds that all translation and learning of sentences, even observation sentences, is dependent on *empathy*. (Ebbs, 1994: 537–538)

I also argued that this radical change in Quine's view of the relationship between stimulus meaning and intersubjectivity occasions a correspondingly radical change in the sense in which Quine's epistemology and semantics are *naturalistic*:

From the point of view of an austere doctrinal naturalism [that equates objectivity with what is settled by the truths of the mature natural sciences], there are no remaining links between intersubjectivity and sensory stimulation. At first this seems to cut *all* ties between intersubjectivity and sensory stimulation, and to set naturalized epistemology adrift. But Quine sees that mentalistic discourse is now needed to describe the relationship between intersubjectivity and sensory stimulation, and so he no longer insists on the austere doctrinal point of view. For example, in §24 of *Pursuit of Truth*, he observes that one speaker's learning of an observation sentence like 'It's raining' is dependent on another speaker's mastery of mentalistic sentences like 'Tom perceives that it's raining'. The learning of observations sentences and of language more generally depends on an 'objective pull' that in effect imposes an intersubjective structure over each speaker's private sensory stimulations. Our understanding of the 'objective pull' is now utterly dependent on our mastery of mentalistic sentences like 'Tom perceives that it's raining'. Thus the link between private sensory stimulation and scientific intersubjectivity is mediated by mentalistic discourse. In this and many other ways, *Pursuit of Truth* presents a fascinating new view of the relationship between naturalistic and mentalistic perspectives on cognitive meaning and objective knowledge. (Ebbs, 1994: 540–541)

2   Quine's reply (first version): 'Preestablished Harmony'

In the first version of his reply, titled 'Preestablished Harmony', Quine generously acknowledges the main points I make in my review, but argues that he can nevertheless explain intersubjectivity naturalistically in terms of what he calls 'the preestablished intersubjective harmony of perceptual similarity standards, rooted in natural selection' ('Preestablished Harmony', paragraph 1, last sentence). He emphasizes that 'my use of the word "empathy" invokes nothing beyond what the mother and the field linguist were already up to according to *Word and Object*. There remains, however, in both the setting of *Word and Object* and that of *Pursuit of Truth*, a question of causal explanation'. The key idea is that

If *A* and *B* jointly witness two events, and *A*'s neural intakes on the two occasions are perceptually similar by *A*'s standards, then *B*'s intakes will tend to be similar by *B*'s. (Quine, 'Preestablished Harmony', four paragraphs from the end)

where

> [This] harmony of innate standards of perceptual similarity is accounted for by natural selection. (Quine, 'Preestablished Harmony', penultimate paragraph, first sentence)

Thus Darwin's work affords 'a naturalistic account of our seeming access to other minds'. (Quine, 'Preestablished Harmony', last paragraph)

3   Quine's reply (first version): 'Response to Gary Ebbs'

In the second version of his reply, titled 'Response to Gary Ebbs', Quine succinctly summarizes the main point of my review: 'Ebbs sees the transition from *Word and Object* (1960) to *Pursuit of Truth* (1990, 1992) as adulterating my naturalism with mentalism' (Quine, 'Response to Gary Ebbs', second paragraph). Against this, he argues that 'Appeal to [empathy] is no breach of naturalism or physicalism, by my lights' (third paragraph, last sentence). One reason he offers in defense of this claim – a reason he does not make explicit in the first version of his reply – is that

> in several passages... Ebbs takes my terms "science" too narrowly. We have no word with the breadth of *Wissenschaft*, but that is what I have in mind. History is as at home in my naturalism as physics and mathematics. So also, indeed, is translation. My conjecture of indeterminacy of translation is just that in the radical translation of theoretical material there may be incompatible alternative turnings and no fact of the matter; either will serve, but not both. (Quine, 'Response to Gary Ebbs', seventh paragraph)

Quine also adds that in a conference in 1986 Davidson, Dreben, and Føllesdal had pressed him on the problem of 'intersubjective homology or near-homology of nerve endings', and it only later dawned on him 'how else to account for continuing agreement over observation sentences' (Quine, 'Response to Gary Ebbs', five paragraphs from the end). He then repeats the statement of harmony of subjective scales of perceptual similarity, namely,

> If A and B jointly witness two events, and A's neural intakes on the two occasions are perceptually similar by A's standards, then B's intakes will likewise tend to be similar by B's. [Note that in the

'Preestablished Harmony' version of this passage, the word 'likewise' was typed in, but later crossed out; in 'Response to Gary Ebbs', the word 'likewise' is reinstated.]

and adds that

> An individual's standards of perceptual similarity can in principle be elicited experimentally by the reinforcement and extinction of responses. [Here Quine cites *Roots of Reference*, 1974: 16–18.] They are largely unlearned, since learning depends on them, but they change gradually with experience. (Quine, 'Response to Gary Ebbs', third paragraph from the end)

He follows this with sketch of how natural selection explains the harmony of innate standards of perceptual similarity. In the final paragraph, there is another small change. In 'Preestablished Harmony' he describes Darwin's revolution as 'the great revolution in metaphysics'; in 'Response to Gary Ebbs' he calls it 'the greatest revolution in metaphysics', with the 'est' written in by hand.

4 Excerpts from my letter to Quine and his reply

Impressed by Quine's hypothesis of a preestablished harmony of innate standards of perceptual similarity, but not fully convinced that Quine's appeal to it addresses my worry that Quine's new account of the intersubjectivity of observation adulterates his naturalism with mentalism, I wrote to Quine as follows:

> I welcome your acceptance of history and translation within science, broadly construed (page 3, first paragraph). But if you accept history and translation as part of science, I don't see how you can distinguish between indeterminacy and underdetermination any more. In your reply to Chomsky you said that "physics" (which I suppose must be read broadly to mean "science") is our ultimate parameter. I interpret you as saying that our best theories of nature are underdetermined, but not indeterminate, since they set the ultimate parameters (subject to change, of course) of our descriptions of reality, whereas the point about indeterminacy is that even once all the naturalistic parameters are set, translation is still indeterminate. But if you are now willing to include our actual translation practices as part of science, why not just accept that these practices set the ultimate parameters for the truth

about translation? Why not settle for a description of what counts in actual practice as a good translation, instead of trying to offer a philosophical reconstruction of what good translations preserve (namely, the totality of speech dispositions)? This question seems particularly pressing, since you grant that "the full or holophrastic indeterminacy of translation draws too broadly on a language to admit of factual illustration" (*Pursuit of Truth*, p. 50). This suggests that in actual practice there is no sign of full or holophrastic indeterminacy, and that it is in fact a purely theoretical possibility, produced and sustained by your reconstruction of the epistemology of translation.

I am also surprised that you now think that appeal to empathy is "no breach of naturalism or physicalism," since in §45 of *Word and Object* you wrote that "the underlying methodology of the idioms of propositional attitude contrasts strikingly with the spirit of objective science at its most representative." (218) The unfavorable contrast between empathy and the spirit of objective science was, I thought, the main reason you did not include the idioms of propositional attitude (even in extensional form) in your canonical scheme for science (*Word and Object*, p. 221). But this double standard can no longer be viewed as a principled expression of your scientific naturalism, if "science" is now to include empathy as one of many methods for the pursuit of truth. It seems the double standard can now be no more than an expression of a purely metaphysical bias for some kinds of scientific descriptions over others.

I also have two small points of clarification about my own claims in the review:

1) By "mentalism" I meant the willingness to *use* and *affirm* such sentences as "Tom believes that it is raining" within the conceptual scheme of science, broadly construed. I can now see that my use of the word "mentalism" was misleading, since someone might take me to be suggesting that you now accept propositions and other intensional entities. I suppose that's why you emphasize that you still "cling to extensionality" (second version, p. 2).

2) On page 3, second paragraph (and briefly also on page 1, third paragraph, second sentence) you suggest that on my reading of chapter two of *Word and Object*, stimulus meanings "were meant to figure in the practice of translation or language teaching." But this is not what I was assuming. As I understood your view in *Word and Object*, whether or not the linguist appeals to (or even knows

of) these facts, they are "the objective reality that the linguist has to probe when he undertakes radical translation" (*Word and Object*, 39). "...he translates...by significant approximation of stimulus meanings." (*Word and Object*, 40). I did not take you to mean that linguists have to know what stimulus meanings are, but just that from a naturalistic point of view, translation of observation sentences is in fact mirrored by (approximate) sameness of stimulus meanings. As I understood your *Word and Object* position, this naturalistic fact (of approximate stimulus meanings) is the "objective reality" that underlies translation of observation sentences. For sentences that are more theoretical, there is no objective reality to be right or wrong about, and so we need to adopt a system of "analytical hypotheses" that go beyond anything implicit in speech dispositions.

In a letter dated August 12, 1995, Quine replied to these comments as follows:

I don't see how my liberal usage of "science", as stretching from quarks to history, bears on the Chomsky issue. It is not new; not a liberalization of *Word and Object* nor of my "Reply to Chomsky." It is not that I accord each branch of science equal footing in our theory of the world. Physics is primary in that any difference between matters of fact must hinge ultimately on difference at the microphysical level, however incapable the physicist or others may be of tracing the dependence. If in any branch of science we find incontrovertible evidence for something (astrology, telepathy) seemingly untraceable to physics even in principle, the physicist has a problem and turns back to the drawing board in hopes of a new breakthrough. The buck stops with him.

Regarding another passage on indeterminacy, you wrote that it "suggests that...holophrastic indeterminacy...is...a purely theoretical possibility." This is right, and from away back, except for the last three words, where I favor "plausible theoretical conjecture."

You liken empathy to the propositional attitudes. I don't. It is the procedure that I ascribed to the translator already in *Word and Object* in behavioral terms, though not yet calling it empathy. The propositional attitudes, on the other hand, are predicates, and ones that even resist classical predicate logic until accommodated (in the *de dicto* case) by quotation and spelling.

Exensionalism has been my uncompromising tenet ever since college days. On other counts I draw no line against mentalistic predicates so long as we recognize philosophically that states of mind are states of body, and methodologically that evidence must admit of articulation in observation sentences. Empathy, unarticulated into observation sentences, would not pass muster as scientific evidence, however indispensable in thinking up hypotheses.

Your interpretation of *Word and Object* on stimulus meanings was right, I now see. The one major departure from *Word and Object* in my later work, so far as I see, is renunciation of intersubjective identity or similarity of stimulus meanings. Preestablished harmony of subjective standards of subjective perceptuality fills the bill. This is a step forward scientifically, for subjective perceptual similarity is testable (*Roots of Reference*) and the harmony is explained by natural selection.

References

Ebbs, G. (1994) 'Review of *Pursuit of Truth*, revised edition, by W.V. Quine', *The Philosophical Review*, 103(3): 535–541.

Quine, W.V. (1992) *Pursuit of Truth*, revised edition (Cambridge, Mass: Harvard University Press).

Quine, W.V. (1996) 'Progress on Two Fronts', *Journal of Philosophy*, 93(4): 159–163.

4
Preestablished Harmony (1995)

W.V. Quine

In my ill-organized way I have only now, a year late, come upon Gary Ebbs' brilliant review of my *Pursuit of Truth*.[1] I am flattered by his scholarly command of my writings and impressed with his penetration to crucial points. I am thankful, amid all this, for his evidently having missed one vital point in my evolving views; for other readers will have missed it *a fortiori*, and I am now alerted to make amends for my inadequate exposition. The point he seems to have missed is the preestablished intersubjective harmony of perceptual similarity standards, rooted in natural selection.

It bears on the learning of observation sentences, by the child in the home language and the field linguist in the alien language. According to *Word and Object*, to learn an observation sentence is to endow it with a 'stimulus meaning' similar to those that it has for other speakers. A stimulus meaning was a set of sets of nerve endings, and I was uneasy even then about intersubjective similarity of such sets. I sketched the field linguist's actual procedure in terms purely of observation of behavior, then as now; the linguist was never meant to know about nerve endings and stimulus meaning. By 1990, I was applying the buzzword 'empathy' to this routine, but it was the same old behavioristic routine.[2]

In 1986, meanwhile, it had tardily dawned on me that since the child and the field linguist equate observation sentences intersubjectively on the basis purely of observable behavior anyway, the privacy of stimulus meanings can be left inviolate, intersubjectively walled off.[3]

If my word 'empathy' hinted of a mentalist in turn, I aggravated the suspicion by writing that the

> handing down of language is ... implemented by a continuing command, tacit at least, of the idiom 'x perceives that p' where 'p'

stands for an observation sentence. Command of this mentalistic notion would seem therefore to be about as old as language. It is remarkable that the bifurcation between physicalistic and mentalistic talk is foreshadowed already at the level of observation sentences, as between 'It is raining' and 'Tom perceives that it is raining'. Man is indeed a forked animal.[4]

It is unwarranted, however, to read this as endorsing mentalism. More than once I have remarked on the serendipitous fruits of confusions.[5]

As for the propositional attitudes, of which I view 'perceives that' as the pioneer, I have long recognized and deplored my inability to get along without them.[6] In this matter I reached my present *modus vivendi* in 1992.[7] As of now I cling to extensionalism, and thus to classical predicate logic as the logic of acceptable scientific discourse. I have reconciled the propositional attitudes *de dicto* with extensionalism, via quotation and spelling, and reconciled myself to banishing propositional attitudes *de re*. These languish in the limbo of auxiliaries, along with the indicator words.

This leaves the mentalistic predicates of propositional attitude aboard, but no breach of extensionality. Such is my commitment to Davidson's anomalous monism. Scientific language thus inclusively conceived tolerates these predicates, albeit as epistemological danglers that neurology and physics can do without. Paul Churchland dreams of reducing them to neurology, and I hardly need say that I should be more than pleased.

Ebbs noted most of the foregoing points in his review. I have now to make my main one. I explained that my use of the word 'empathy' invokes nothing beyond what the mother and the field linguist were already up to according to *Word and Object*. There remains, however, in both the setting of *Word and Object* and that of *Pursuit of Truth*, a question of causal explanation.

Take the case of the mother and child; the other case is parallel. *Word and Object* has the mother inducing in the child a stimulus meaning for the observation sentence 'Milk' similar to her own. Similar stimulus meanings cause similar responses to similar stimuli, so the child's subsequent use of the observation sentence agrees with the mother's. But this causal explanation appeals to intersubjective similarity of stimulus meanings.

In *Pursuit of Truth* that appeal is out of order, so the causal question recurs: why, after the mother has got the child to affirm 'Milk' once in an appropriate situation, does the child's usage continue to agree with the mother's? The answer can lie no longer in intersubjective similarity of stimulus meaning. It now lies rather in an intersubjective parallelism of subjective scales of perceptual similarity. If *A* and *B* jointly witness two events, and *A*'s neural intakes on the two occasions are perceptually similar by *A*'s standards, then *B*'s intakes will tend to be similar by *B*'s.

Intersubjective harmony of similarity standards is needed not only in accounting one presentation of milk similar to another, but also in accounting one utterance of 'Milk' similar to another. Failing such harmony, the child's continuing heralding of milk would fall on uncomprehending ears.

The harmony of innate standards of perceptual similarity is accounted for by natural selection. Similarity is the basis of expectation, for we have an innate tendency to expect similar events to have sequels similar to each other. This is primitive induction. Accordingly a scale of perceptual similarity has survival value insofar as it is conducive to successful expectation, and hence to anticipation of predator, prey, and other threats and boons. Shared environment down the generations would make, then, for parallel similarity scales. Difference of environment would make eventually for difference in similarity scales between different peoples, but the lot of humanity around the world down the ages has been enough alike to make for parallelism of evolution in its main lines.

Darwin wrought the great revolution in metaphysics, out-distancing Copernicus; for he reduced final cause to efficient cause. Now we see his theory at work in epistemology, affording a naturalistic account of our seeming access to other minds.

Notes

1. *Philosophical Review* 103 (1994), pp. 535–541.
2. *Pursuit of Truth* (Cambridge: Harvard, 1990), p. 42.
3. *Ibid.*, p. 41.
4. *Ibid.*, pp. 61–62.
5. E.g. in *Roots of Reference* (LaSalle, Ill.: Open Court, 1974), pp. 68, 125.
6. Thus my 'double standard' in *Word and Object*, pp. 216–221.
7. *Pursuit of Truth*, 2nd ed. (1992), pp. 65–72.

Bibliography

Ebbs, G. (1994) '*Pursuit of Truth*, by W.V. Quine', *Philosophical Review*, 103: 535–541.

Quine, W.V. (1960) *Word and Object* (Cambridge, Mass: MIT Press).
Quine, W.V. (1974) *Roots of Reference* (LaSalle, Ill.: Open Court).
Quine, W.V. (1990) *Pursuit of Truth* (Cambridge: Harvard).
Quine, W.V. (1992) *Pursuit of Truth*, second edition (Cambridge: Harvard).

5
Response to Gary Ebbs (1995)

W.V. Quine

In my ill-organized way I have only now, a year late, come upon Gary Ebbs' brilliant review of my *Pursuit of Truth*.[1] I am flattered by his scholarly command of my writings and impressed with his penetration to crucial points. I am thankful, amid all this, for his evidently having missed vital points in my evolving views; for other readers will have missed them *a fortiori*, and I am now alerted to make amends for my inadequate exposition.

Ebbs sees the transition from *Word and Object* (1960) to *Pursuit of Truth* (1990, 1992) as adulterating my naturalism with mentalism. He infers this from my abandonment of the intersubjective matching of stimulus meanings and my resort to empathy.

'Empathy' has a mentalistic ring, but the procedures that it denotes in *Pursuit of Truth* (p. 42) are just the observations of behavior that I had ascribed to the field linguist in *Word and Object* (pp. 29–30). The linguist was never meant to have any conception of stimulus meanings or of the nerve endings of which they are composed. The notion of stimulus meaning does not even emerge until a later page (p. 32). Empathy, as it figures thus anonymously in *Word and Object* and by name in *Pursuit of Truth*, is mentalistic only in the negative sense of being manifested in behavior rather than defined physiologically. Appeal to it is no breach of naturalism or physicalism, by my lights.

If my word 'empathy' hinted of a mentalistic turn, I aggravated the suspicion by writing that the

> handing down of language is ... implemented by a continuing command, tacit at least, of the idiom *"x perceives that p"* where *"p"* stands for an observation sentence. Command of this mentalistic notion would seem therefore to be about as old as language. It is

remarkable that the bifurcation between physicalistic and mentalistic talk is foreshadowed already at the level of observation sentences, as between "It is raining" and "Tom perceives that it is raining". Man is indeed a forked animal. (*Pursuit of Truth*, pp. 61–62)

It is unwarranted, however, to read this as *endorsing* mentalism. More than once I have remarked on the serendipitous fruits of confusions.[2]

As for the propositional attitudes, of which I view 'perceives that' as the pioneer, I have long recognized and deplored my inability to get along without them.[3] In this matter I reached my present *modus vivendi* in 1992.[4] As of now I cling to extensionalism, and thus to classical predicate logic as the logic of acceptable scientific discourse. I have reconciled the propositional attitudes *de dicto* with extensionalism, via quotation and spelling, and reconciled myself to banishing propositional attitudes *de re*. These now languish in the limbo of auxiliaries, along with the indicator words.

This leaves the mentalistic predicates of propositional attitude aboard, but no breach of extensionality. Such is my commitment to Davidson's anomalous monism. Scientific language thus inclusively conceived tolerates these predicates, albeit as epistemological danglers that neurology and physics can do without. Paul Churchland dreams of reducing them to neurology, and I hardly need say that I should be more than pleased.

It seems in several passages that Ebbs takes my term 'science' too narrowly. We have no word with the breadth of *Wissenschaft*, but that is what I have in mind. History is as at home in my naturalism as physics and mathematics. So also, indeed, is translation. My conjecture of indeterminacy of translation is just that in the radical translation of theoretical material there may be incompatible alternative turnings and no fact of the matter: either will serve, but not both. This is a caveat regarding the notion of meaning, and not to be read as *traduttori traditori*.

Ebbs sees my privatization of stimulus meanings, in passing from *Word and Object* to *Pursuit of Truth*, as more drastic than I do. Stimulus meanings still figure in *Pursuit*, and they never were meant to figure in the practice of translation or language teaching. Their role early and late was causal: they are the neural launching pads of observation sentences.

Intersubjective likeness of stimulus meanings served, in its day, to account causally for our continuing intersubjective agreement in the affirmation of an observation sentence from occasion to occasion.

The child's usage and the mother's, or linguist's and the native's, do not drift apart along divergent paths of extrapolation. Take the case of mother and child; the other case is parallel. *Word and Object* has the mother inducing in the child a stimulus meaning for the observation sentence 'Milk' similar to her own. Similar stimulus meanings cause similar responses to similar stimuli, so the child's subsequent use of the observation sentence continues to agree with the mother's. The stimulus meaning is a set of sets of receptors, and the relative stability of the receptors would arrest drift. If the child's stimulus meaning of the observation sentence matched the mother's (or the linguist's the native's) on the first occasion, it will continue to.

Still, even at the time of *Word and Object*, I was uneasy about this intersubjective matching of stimulus meanings. It called for an intersubjective homology or near-homology of nerve endings that I felt ought to be irrelevant. Davidson, Dreben, and Føllesdal pressed the problem at our little conference in 1986. At length it dawned on me how *else* to account for continuing agreement over observation sentences.

It is due rather to a preestablished harmony of subjective scales of perceptual similarity. If *A* and *B* jointly witness two events, and *A*'s neural intakes on the two occasions are perceptually similar by *A*'s standards, then *B*'s intakes will likewise tend to be similar by *B*'s.

An individual's standards of perceptual similarity can in principle be elicited experimentally by the reinforcement and extinction of responses.[5] They are largely unlearned, since learning depends on them; but they change gradually with experience.

The harmony of innate standards of perceptual similarity is accounted for by natural selection. Similarity is the basis of expectation, for we have an innate tendency to expect similar events to have sequels similar to each other. This is primitive induction. Accordingly a scale of perceptual similarity has survival value insofar as it is conducive to successful expectation, and hence to anticipation of predator, prey, and other threats and boons. Shared environment down the generations would make, then, for parallel similarity scales. Difference of environment would make eventually for difference in similarity scales between different peoples, but the lot of humanity around the world down the ages has been enough alike to make for parallelism of evolution in its main lines.

Darwin wrought the greatest revolution in metaphysics, out-distancing Copernicus; for he reduced final cause to efficient cause. Now we see his theory at work in epistemology, affording a naturalistic account of our seeming access to other minds.

Notes

1. *Philosophical Review* 103 (1994), pp. 535–541.
2. E.g. in *Roots of Reference* (LaSalle, Ill.: Open Court, 1974), pp. 68, 125.
3. Thus my 'double standard', in *Word and Object*, pp. 216–221.
4. *Pursuit of Truth*, 2nd ed., pp. 65–72.
5. See *Roots of Reference*, pp. 16–18. In the present paper, I am indebted to Burton Dreben for helpful discussion.

Bibliography

Ebbs, G. (1994) *'Pursuit of Truth*, by W.V. Quine', *Philosophical Review*, 103: 535–541.
Quine, W.V. (1960) *Word and Object* (Cambridge, Mass: MIT Press).
Quine, W.V. (1974) *Roots of Reference* (LaSalle, Ill.: Open Court).
Quine, W.V. (1992) *Pursuit of Truth*, second edition (Cambridge: Harvard).

Part II

Quine's Contact with the Unity of Science Movement: A Glimpse of His Friendship with Ed Haskell

6

Observations on the Contribution of W.V. Quine to Unified Science Theory

Ann Lodge, Rolfe A. Leary and Douglas B. Quine

The position of Quine concerning scientific rigor in philosophy has been well established. This paper is concerned with his influence upon development of unified science theory, largely in terms of models put forward by Edward Froehlich Haskell. This discussion will be largely drawn from the correspondence of these two men and a few other colleagues in order to let them speak for themselves in their own words. Harold Cassidy, former Professor Emeritus of Chemistry at Yale, was intimately involved in these exchanges. The available correspondence* began in the late 1930s and spanned roughly five decades.

Ed was Van's best friend, as he affirms in his autobiography, *The Time of My Life*. Van was also probably, fortunately, Ed's severest critic. The men's friendship began at Oberlin College, Ohio, where they were undergraduates in the late 1920s and lasted throughout their lifetimes. Their relationship began when Ed moved into an off-campus house dubbed Arthron (the 'Joint') where Van already lived. Ed wrote 'college changed wonderfully for me. The campus of frustrating separate departments became a boundless interdepartmental forum' (Haskell, 1980: 37). This forum included, among others, Van Quine in philosophy and math, Fred Cassidy in English and philology, his brother Harold Cassidy in chemistry and biology, and G. Townsend Lodge in psychology. Representing a cross-section of the four major university divisions, the physical, biological, psycho-social-political sciences and the humanities, the group regularly had intense interdisciplinary bull sessions in the attempt to reach a degree of mutual clarity.

Ed writes, 'Hindsight now shows that we were preparing to synthesize them to assemble them into *a* single discipline'. He goes on to describe

Figure 6.1   Edward F. Haskell, Fred G. Cassidy, W.V.O. Quine, and Harold G. Cassidy on front porch of off-campus house in which they lived 1926–1930. Photo taken at 50th class reunion Oberlin College, 1980

Source: Photo credit: *Oberlin Alumni Magazine*.

Van Quine as 'our leading light ... not an empirical scientist but a logician, a formal scientist. Since it is formal structure that diverse-appearing phenomena – physical, biological, mental – have in common, and since our bull sessions could never be dominated by any single empirical discipline, they were guided by the formal rules that all of the sciences and humanities have in common: the basis of their common unifying discipline' (Haskell, 1980: 37–38). These men were to continue to work together and influence each other theoretically, albeit from a distance, for much of their lives.

In June, 1939, Van wrote Ed as follows: 'The program put forward in your "Mathematical Systematization of 'Environment', 'Organism', and 'Habitat'" (Haskell, 1940) strikes me as highly important. A unified theory exhibiting life and entropy as opposed and balanced tendencies would constitute a fundamental and far-reaching synthesis. And the plan of construing all natural entities explicitly as processes, disposed in space-time according to equations, should make for essential clarifications of ecology and render that field amenable to powerful techniques'. Van then goes on to raise questions concerning Ed's definitions, stating 'I am more rigorous than most in insisting on careful distinction between "words" and designata' (WVQ to EFH, June 24, 1939). The question also arises as to whether Ed may have used the word 'ecology' in place of 'system' or 'ecosystem'. Today the systems ontological perspective seems dominant in many different disciplines.

In a subsequent letter, Van writes,

> I have made almost no comments on the margins of your manuscript; for I feel that I agree with just about every statement, though in many cases this agreement is preceded by a certain amount of conjectural interpretation. You recognize of course that much must be done in the way of rigorization before the affirmed relationships can become definitely established...A general question arises in connection with your thesis of the all-inclusiveness of ecology. Sciences are distinguished from one another not merely by the arrays of individuals to which they apply, but by the differing conceptual schemes whereby they correlate those individuals. (WVQ to EFH, July 22, 1939)

In his autobiography, *The Time of My Life*, Quine wrote:

> On the way back from an Akron Christmas in 1945 I had a long session with Ed Haskell and Harold Cassidy in New York. Harold was by then a chemistry professor at Yale. Ed's communist fervor had been reversed by his intimate acquaintance with the party and the system. He was now taken up with broad ideas about cooperation and predation. There were some statistical findings regarding fish populations in which he saw implications for human societies. He was well informed on social and political matters and much concerned about them. Within this compass his ideas were good and capable of bettering society. He was now venturing to generalize them on a cosmic scale, however, into something that he called "unified science"; and I tried to apply the brakes of rigor to his runaway ambition. Our meeting was followed up by an eighteen-page letter from him and a twenty-page response from me. (Quine, 1985: 191–192)

The letter from Ed Haskell, referred to above by Quine, seemed to express his opinion that modern logic was dissociated from empirical science, while the latter remained in disarray for lack of a unifying system of classification. He literally pleaded with Quine to undertake 'a full, clear

facing of the basic philosophical problem, the statement of the alternatives, the deliberate selection of one which conforms to the spirit of
science, and the presentation of a consistent and complete theory of
classification based thereon would clear the whole road which we must
rapidly traverse if we are to head off world catastrophe. How about it!'
(EFH to WVQ, December 24, 1945). Quine replies:

> Perhaps you say then that I should change fields; that my present
> field (call it what I may, is staked out in my Introductions) is too
> remote from the atomic-bomb problems. This reminds me of an
> evening in Washington when Wundheiler, a Pole, tried to talk Tarski
> and me into becoming physicists and directing our "clear thinking"
> upon what Wundheiler takes to be confusions in thermodynamics.

He goes on to point out that during World War II,

> 'I abandoned theoretical studies of my choice in order to devote my
> full energies to the prosecution of the war. This I continued to do for
> more than three years (winning a letter of commendation from my
> admiral). Having paid this voluntary tithe, I'm going back to the field
> where my chief interests and talents, if any, lie, with the idea that it
> too had its importance'.

He further comments,

> I should add: I see no field for real accomplishment in the matter of
> general theory of classification; I see no reason to believe that there
> are any useful "general" principles here even awaiting discovery (i.e.,
> apart from specific systems of classification as in sociology), beyond
> the superficial sort of common-sense advice in the old books such
> as Jenson's; and I see no reason (as I think I remarked to you in New
> York) to consider the absence of such a theory the bottleneck of soci-
> ology ... Sociology "can" probably be given good classification, and it
> may even be fruitful to draw on analogies in chemistry and biology
> for occasional inspiration ... But there is no reason to believe in the
> possibility of a substantial "general" theory of classifications valu-
> able to all science. (WVQ to EFH, January 4, 1946)

Thus, it appears that a clear difference of opinion evolved between
the two as to how far a unifying approach to science should or could
meaningfully proceed. Nevertheless, the discussion and the friendship
continued as productively as ever.

   Haskell continued his efforts to develop an interaction geometry that
demonstrated usefulness in the analysis and synthesis of data in a wide

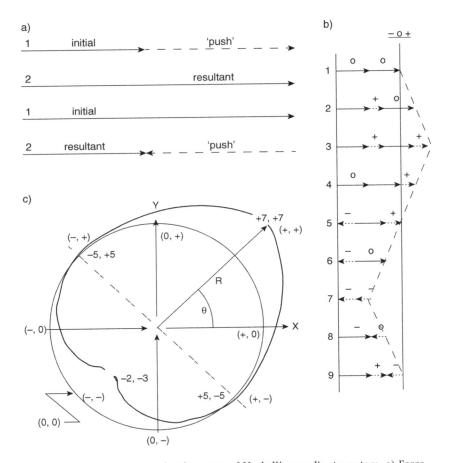

Figure 6.2 Three stages in development of Haskell's coordinate system. a) Force vectors, adapted from Einstein and Infeld (1938). These two diagrams apparently provided the spark for Haskell's work in the 1940s. b) Completion of all possible interaction vector diagrams (*Main Currents in Modern Thought*, 1949). The interaction numbers have been added, and also the dashed line that connects arrow lengths. Note interaction 8 is mislabeled (should be 0,-). There are 9 qualitatively different types of interaction. c) Transformation of linear array of interactions in b) into a mathematical coordinate system – initially called the Periodic coordinate system – that shows both type and intensity of interaction. The dash-dot line from lower right to upper left identifies where the 'gain' and 'loss' from interaction are offset (hence the '=' symbol) and where the intensity of interaction is zero, hence the 'O' symbol. This they called the Zero-equal axis, which dominated much early correspondence among Haskell, Quine and Cassidy (Source: Harold Cassidy's unpublished book about Ed's work). The curve in c) is known as the coaction cardioid, and separates the 'conflictor's deficit' (lower left) from the 'cooperator's surplus' (upper right). Haskell's coordinate system nicely separates interaction type (θ) and interaction intensity (R). Dindal (1975) proposed a form of Haskell's classification grid that also incorporated interaction intensity by adding more cells to a grid.

variety of scientific areas. He derived this approach from a glimpse of pattern he saw in Einstein and Infeld's diagrams in their 1938 book, *The Evolution of Physics*, which he first systematized into a categorizing ('pigeon holing') scheme and then into a mathematical coordinate system (Figure. 6.2). Part of Ed's genius was his ability to detect patterns, or the possibility of patterns, in minimal samples of information. For example, Einstein and Infeld had given illustrations of just two interaction diagrams (Figure 6.2a), but Ed saw there are logically nine different types of interaction, Figure 6.2b)

Somehow, Haskell wondered if the Periodic coordinate system wasn't a synthesis of the three geometries – Euclidean (zero-zero circle), Lobachevski (lower left, between (0,0) circle and coaction cardioid), and Riemannian (upper right, between (0,0) circle and coaction cardioid). Efforts by at least two mathematicians proved inconclusive on the question. Following a heavy night of Arthron poker in 1983, Ed wrote to Van and wondered if the (O=) axis might be relevant to 'Quine's Theory of Values' (EFH to WVQ, October 14, 1983). The authors have been unable to find materials suggesting Quine had actually articulated a 'Theory of Values'. Arthron poker matches were much talked about in correspondence – 100 chips, worth 1 cent each – with much heavy betting and cumulative losses.

Haskell saw what he thought were important population patterns in studies of fish in Ohio fish ponds, based on published research (Langlois, 1936). The population patterns reminded him of Mendeleev's periodic table of the chemical elements. Haskell's correspondence files contain a letter to Langlois commenting on another scientist's questioning of Ed's use of Langlois' data. Did Ed change Langlois' data to fit his theory? Ed says he did not; 'I think that you will see that I have altered none of your findings, merely questioned the correctness of certain ones' (EFH to Langlois, June 12, 1948). The two seemed to be on cordial terms in future correspondence.[1] Discovering periodicity in Langlois' data led Ed to generalize the periodic table concept in his attempt to help 'unify' all sciences. Interestingly, Schwemmler (1984) presents a system of 'reconstruction of cell evolution: a periodic system' that references Haskell, and proposes periodic tables of cell evolution.

Van sent Ed a lengthy critique of his geometric models in the following letter excerpted below (WVQ to EFH, June 20, 1955):

> This three-speed space described in your paper is quite a dish. I can't claim to understand all you say about it. But I hope that you are right in believing it to provide the required geometrical representation for

your empirical theory, and that other parts of the book may clear up any really essential obscurities.

A misconception regarding mathematics and mathematicians recurs in your paper, and I must try and talk you out of it. You suggest that mathematicians are traditionalistic, and that you have had a certain advantage of freedom through being, as you say, mathematically semi-literate. Nothing could be farther from the facts. Pure mathematicians are the breed, of all breeds, most eager and able to reverse a plausible postulate and try the consequences. This, and the quest of generality upon generality without regard to past practice or possible purpose, are traits overwhelmingly present both in the history of mathematics over the past century and in individual research mathematicians as I have known them. In mathematics revolution is routine, and without fanfare. Unprecedented departures, based on the most tenuous of analogies, turn up for a page or two and are superceded by further ones in the next.

Geometry, even in infinitely many dimensions, Euclid and non-Euclidean, continuous and discontinuous – all this is trodden ground. If you were to set forth a geometry in infinitely many dimensions, Riemannian in some dimensions and Euclidean or Lobachevskian in others, no mathematician would raise an eyebrow. "All right, let's see what you are doing" – that would be the typical response. The specific piece of work might or might not be found elegant, surprising, suggestive.

In point of fact no mathematician would bat an eye over a hybrid space containing Euclidean, Riemannian, and Lobachevskian regions. It is only a question of zero, positive, and negative curvature of space; and you could very well have a space with different curvatures in different parts. This situation is easily actualized in two-dimensional space, thought of as a bent surface in the familiar three-dimensional Euclidean world. Just picture part of it as a fist, part ellipsoidal and part as hyper-coloidal. Afterward we carry the analogy to more than two dimensions.

Einstein's own theory posits a four-dimensional space with varying curvatures. People call it Riemannian because the curvatures, though varying, tend to be positive; but there is in principle as much difference between one positive curvature and another and another as there is between positive and zero (Riemannian) or negative (Lobachevskian). For that matter, any Riemannian or Lobachevskian space is already, in principle, Euclidean in the small.

A portion of Ed's response is presented below (EFH to WVQ, June 28, 1955):

> And thanks for the very enlightening letter. I hasten to alter my paper at the points where you object. To wit: I am clarifying passages which give rise to the impression that I think fear of inconsistency deterred mathematicians from this synthesis ...
>
> I am delighted with your statement that "Any Riemannian or Lobachevskian space is already, in principle, Euclidean in the small." For some years now, Townsend and I have been very successfully using a slightly modified cartesian coord. syst. wherever phenomena (psychological, biological or other) belonging to a single Period are being represented. I have been justifying this success by asserting that Euclidean geometry applies within any given Period. This assertion conforms to your 'in the small,' and is thus correct.

Haskell published: 'Mathematical Systematization of "Environment", "Organism", and "Habitat"' in *Ecology* in 1940, 'A Natural Classification of Societies' in the *Transactions of The New York Academy of Sciences, Series II*, in 1947, followed by 'A Clarification of Social Sciences', in *Main Currents in Modern Thought* (including cover illustration) in 1949. In 1948, he was invited to organize a three-day interdisciplinary symposium, *Cooperation and Conflict Among Living Organisms*, for the annual meeting of the American Association for the Advancement of Science (AAAS). Included were a number of distinguished scientists as participants, including Paul Burkholder (Yale), Paul Sears (Oberlin), James Bonner (CalTech), Robertson Pratt (U. California), Laura Thompson (Institute of Ethnic Affairs) and others. The manuscripts for these talks are still in files of Haskell's materials; however, we have been unable to locate a resultant record or publication. At the conclusion of the conference, Haskell was joined by several other scientists to form a small 'think tank' called Council for Unified Research and Education (CURE), which he headed for almost 40 years.

Harold Cassidy sent the following commentary, to both Van and Ed, as follows (HGC to EFH and WVQ, January 18, 1946):

> I was quite taken with E's vision of a general approach to all systems of classification. That idea has occurred to me in a less specific form as a sort of utopian extrapolation from the special classifications. However it is a hell of an extrapolation, and in my more sober moments I doubt whether, as Van puts it "there are any useful general

principles here even awaiting discovery" (p.8). The reason I doubt this lies in the implications of the word "useful". It is certainly so that a certain amount of classification is useful – in fact my course in methods of organic chemistry classification (really methodology) has gradually molded itself into a search for suitable classification procedures for processes and operations – however I have observed that if generalization is carried too far one reaches a stage of, to use the economist's term, diminishing returns. This may proceed, in extreme cases, to a state actually of increasing inutility. In the grasping for a unity which is aesthetically satisfying one loses practical utility. This seems to me to be a natural phenomenon in the way of dilemmas.

In a letter dated April 4, 1954, Harold took Van to task for the severity of his criticism of Ed during a particularly difficult and frustrating period in Ed's work. Van's reply (April 9, 1954) agreed that 'my recent blast was unfortunate', commenting also,

> You know, more vividly than I, how friendship and science can conflict. Here is a close friend whose ideas are his life. Up to a point you can serve science and him by arguing him out of his worst ideas and into some better ones. But when he can't be dissuaded from an idea which your scientific standards compel you to reject, then the fundamental conflict sets in between the friendly determination to see him prosper and the scientific determination to see truth prevail.

He went on to explain that he had come to feel that 'kindness had long been getting the better of candor...It was best, I decided, to treat him as one expects to be treated: as a tough and responsible scientific mind, under unlimited liability on every published point'. Van further acknowledged that Harold had 'fought this thing through' by collaborating intensively with Ed on his manuscripts. He wrote,

> I have stayed on the sidelines, leaving my good wishes for Eth [*Ed's Arthron name*] relatively unimplemented, and having an easier time of it. Talking with him and reading his releases over the years, I have combined friendly encouragement with objective criticism in a comfortable sort of way, turning to other things when I reached an impasse. (WVQ to HGC, April 9, 1954)

One of us (RAL) read the 'blast' letter in files in the home of Harold Cassidy in 1991. Unfortunately, the Cassidy correspondence may have been lost.

[Van Quine kept all correspondence – incoming and outgoing. For example, Van sent Haskell all the letters he had received from Haskell, beginning in 1929 – for assistance in writing his autobiography. Same with Cassidy – Quine sent Cassidy all correspondence Cassidy had sent him.]

During the mid to late 1940s period, and continuing into the 1980s, both Van Quine and Harold Cassidy maintained fairly intensive three-way communication, which included encouragement, support and often very specific criticism and pruning suggestions concerning Ed's evolving work. A recurring theme in the criticism of both was Ed's tendency, at times, to use abstruse language, to portray the work of others as limited, less visionary, and to claim what they often saw as over-inclusiveness with regard to his own theory. Ed generally took these criticisms to heart and modified his writings as a result.

Harold described Ed as 'one of the most original thinkers I know' (HGC to Dr. E. G. Mesthene, September 2, 1965). Van wrote to Ed:

> What I keep wishing you would do is produce a direct, unprefaced, unpretentious exposition of what strikes you as the newly uncovered truth about coaction, predation, atoms, etc., in the simplest practicable language, using only the irreducible minimum of unfamiliar terms. Even when a new term is needed for clarity, it should be as unstartling and familiar in form as possible. Given the knowledge or ideas about the world that you want the reader to grasp, the rest is after all, a teaching problem. (WVQ to EFH, February 13, 1967)

Haskell worked for years to bring his developing book, originally titled *An Introduction to Unified Science*, and later, *Unified Science: Assembly of the Sciences into a Single Discipline*, to fruition and publication. In 1972, a book which presented much of his theory, as well as applications developed by others, was published under his editorship (Haskell, et al., 1972). [The entire text of *Full Circle* is downloadable at Timothy Wilken's website (see references).]

Quine made the following comments in a letter to Ed preceding publication of *Full Circle*:

> I have no reservations about your coaction theory at the social or normative end. When we come to your parallels of coaction in chemistry and physics, I react with interest but not with enthusiasm that a more explanatory synthesis would arouse. My point here is that I see no mechanism to account for the parallels. I do realize that explanatory mechanism has finally to peter out into description; witness gravitation or the speed of light. Still, explanatory mechanism is great

stuff as far as it goes, and the more of it the better ... Hence some reservations over your way of unifying science. Unification by unity of explanatory mechanism would be great ... Methodological unification immediately recommends itself too. But your unity by parallel structure is rather something for me to ponder and wonder at, hoping we may sometime see how come. (WVQ to EFH, September 22, 1969)

It would seem Haskell had chosen a route to universal explanation quite different from the one suggested by Quine (Figure 6.3). Exactly how Haskell planned to reach the lower right hand corner was never stated explicitly. However, from the methods Haskell used it appeared he hoped to apply a systems ontological perspective at a high level of abstraction, and to look to the cybernetic concepts of system 'work component' and 'controller component', and to the relation between the two summarized in what Ed called the Periodic Law:

$$R = f(\theta)$$

where, R designates system properties proportional to the length of the radius vector, and θ designates the interaction between each system's controller and work components (Figure. 6.2c).

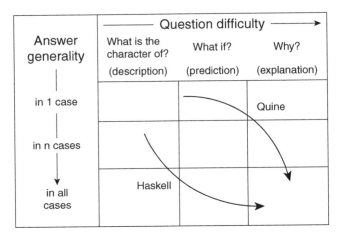

Figure 6.3 Relation between question difficulty and answer generality (Leary 1984b, 1985a). Young scientists often begin their careers in the upper left cell. Few work as long, intensively, and make as many personal sacrifices, to move toward the lower right cell as Ed Haskell.

In the 1960s, Haskell had made contact with Prof. Jere Clark, Southern Connecticut State College, New Haven, CT, who shared many of his interests in general systems theory, analysis, and synthesis. Ed was invited to teach courses on 'unified science' at SCSC that, it turned out, appealed to followers of Rev. Sun Myung Moon. These students introduced Ed to Rev. Moon, who underwrote the expenses for publication of *Full Circle*, and also an international conference with the theme: 'Moral Orientation of the Sciences' (conference flyer, November 23–26, 1972). Displayed prominently on the flier cover is a stylized Haskellian (at the time called 'Periodic') coordinate system with coaction cardioid (see Figure 6.2c). About 14 papers were delivered over the three days, with discussion and comment opportunities.

Quine presented a paper at the Conference, after consultation with Haskell to the effect 'What do you want me to talk about?'. Ed's response: 'May I suggest as the possible topic of your Chairman's remarks: "The Hierarchy of Abstraction Ceilings"?' (EFH to WVQ, October 12, 1972). Also presented were Harold Cassidy's 'Essential Ideas in Unified Science' and G. Townsend Lodge's paper 'The Hierarchy of Selves and Their Coactions'. As had happened before in Haskell's career, recall 1948 symposium at AAAS meeting, the papers delivered at the 1972 Moon conference were never published. In his autobiography afterwards (pp. 395–396), Quine recalled the conference:

> In November 1972 I held forth under the improbable joint auspices of Ed Haskell and Sun Myung Moon. Ed had fallen in with Moonies and had admired their attitudes, notable their anti-communism. The contact had led to Moon's being impressed by Ed's ambitious theory of unified science, to the point of underwriting a First International Congress thereof, held at the Waldorf Astoria. Out of friendship I contributed a slight paper on hierarchical structures, this being a topic that figures in Ed's theory but admits of neutral treatment. Fred and Harold Cassidy came. Harold, indeed, had been working earnestly with Ed down the years, bringing scientific knowledge and restraint to bear. The Congress was to be an annual event; Ed's endeavors over the years were to be crowned with success at last. He had hit the big time, but not for long. Moon was advised against letting Ed's theories dominate the Second Congress, held in Japan the next year. Ed participated in it and then broke with Moon. The congresses have continued without him.

As a college senior aware of the media attention associated with the Moonies, co-author DBQ attended the First Congress as a news reporter representing the Princeton University radio station, WPRB-FM. Later, in 'The Time of My Life' (p. 416), Quine also wrote about the 1975 event:

> A week later Marge and I were at the Waldorf-Astoria for Sun Myung Moon's Fourth International Congress for Unified Science. Ed Haskell was no longer involved. I had declined the previous one, in London, as well as the one in Tokyo; but this time I accepted, for my duty was limited to commentary, there were distinguished participants, the trip was easy, and it could be a good show. I was on a panel ably handled by Ernan McMullen, a Jesuit philosopher whom I had known at Notre Dame. The final event was a dry but otherwise elaborate banquet at which we were entertained by a male chorus and a troupe of Korean ballerinas, both excellent. Then Moon spoke, amusingly at first, in somewhat halting English. "You have had the entertainment; now comes that damned commercial" – such was the gist. "Don't think I'm going to sing to you. Or, on second thought, perhaps I will." And he launched into a Korean song. Voices all around the great dining hall took up the chorus. Presently, however, the affair deteriorated. Moon droned on for perhaps an hour in Korean, through an interpreter. It was a sermon at the intellectual level of a fundamentalist revival meeting. I looked at the dignitaries flanking him at the high table – men who had given the keynote papers and organized the panels. Eugene Wigner, Alvin Weinberg, Sir John Eccles, and other Nobel laureates were among them. As well-behaved guests they listened respectfully to the persistent insult to their intelligence. At last old Wigner got up and hobbled off. Inwardly I applauded this gesture of self-respect, but too soon; through with his private errand, he hobbled back.

During the 1970s and 1980s, Harold worked especially intensively with Ed to help "dejargonize" his theoretical work and make it more comprehensible and publishable. A (Haskell and Cassidy, ~1977) 181-page typewritten manuscript was discovered by co-author DBQ at Quine's old summer cottage in October, 2014; the manuscript date is estimated by the extensive citations which end with a single reference from 1977. Subsequently, Harold, with the help of his wife, Kathryn, finally did manage to come up with a final draft of 1020 pages, which represented their joint work as well (Haskell and Cassidy, unpublished manuscript). This was during the time in Ed's life when he was struggling with his final illness. Ed died on May 5, 1990.

In a letter to Marge and Van Quine (June 11,1989), Harold expressed the following concerns:

What seems to be happening today is that positivistic science carried to an extreme has developed an "epistemological systematic" which does not allow for those "free inventions of the human spirit" [*Einstein*] that can open up vast new domains of "unutilized potential" which are necessary if science is not to become stultified, and is to grow deeply.

In his reply (October 1, 1989), Van wrote,

I grieve as you do over Ed. During the fifty-odd years since Oberlin, he was my most frequent company outside the family. I counted on his quick and novel insights, his inquiring mind, his familiarity with languages and mores, and his congenial sense of the ridiculous. [Reacting to Van's extended travels and circuitous routes between a US East Coast city to a West Coast city *via* Adelaide and Tokyo, Ed once offered a new geometric construct – the greatest distance between two points is 'as the Quine flies'. (EFH to WVQ, June 30, 1959)]

Van further commented eloquently,

So, while I agree regarding the paralyzing effect of positivism as I am understanding it, I don't share your gloomy view of science as caught in positivism's toils. I am awestruck by the dazzling breakthroughs of science, ever more frequent on so many fronts. Ever more particles and galaxies. Close range surveys of remote planets. Newly isolated viruses; mitochondria; new antibiotics. Startling revelations of the devious cerebral mechanisms of vision and of habit formation, giving promise of clearing up the mysteries of memory and cognition...On the other hand I agree that the moral 'sciences' are a dead loss, and we are in a desperate fix. Overpopulation. Pollution and destruction of the environment. Drugs. Crime. Terrorism. For all the glories of natural science, which I have just applauded, our world is hurtling into its apocalypse. This is what has exercised you and Ed and others of the sane minority.

Although Quine expresses doubt about the utility of Haskell's classification scheme for the 'moral sciences', Ed continued to make creative applications of his ideas – unfortunately mostly unpublished. Of particular interest, however, is a two-page summary of a talk 'Control of Power by Values' (delivered at the Men's Club of Riverside Church, New York City, October 26, 1960),[2] which would seem to have immediate application to the rich – poor divide in many nations currently.

Diverse scientists would probably agree that the universe is an organized working system. Niels Bohr states, 'Not withstanding the admittedly practical necessity for most scientists to concentrate their efforts in special fields of research, science is, according to its aim of enlarging human understanding, essentially a unity' (Bohr, 1955). Overall, the conglomeration of specialized concepts and nomenclature that have accrued has made intercommunication, and consequently, productive systematic interaction, among disciplines difficult, if not impossible. A small number of scientists have attempted to bridge the gap by introducing a more synergistic approach. Ed Haskell was an outstanding example.

There have been various approaches to achieving more unified science. Until a few hundred years ago, science, such as it was, existed in a unified state, as part of philosophy. Then increasing scientific discoveries began to lead away from philosophy into specialized subject areas which could be examined empirically through scientific method. As early as 1934, plans were developed for a series of annual congresses on unity of science at a congress at Charles University in Prague (Neurath, 1955). The encyclopedic approach, exemplified and developed in these volumes, represent an attempt at integration through such shared commonalities as methodology, mathematics, logic and semantics. Another approach based primarily on organizational and mathematical principles which are postulated to apply across scientific disciplines is exemplified in the work of Ed Haskell, which has been described.

The correspondence between Van Quine and Ed Haskell would seem to indicate that Van exerted a crucial moderating influence on Ed's theoretical work, especially in terms of focus and rigor. We also are quite confident that Quine helped to increase Ed's awareness of related historical, philosophical precedents, such as J. S. Mills' natural classification, Leibniz' universal characteristic, and other relevant concepts. Perhaps one of the most important points Van asserted to Ed was that unification must happen at the level of explanation rather than description. Ed came up with remarkable interrelated descriptive data in several areas and was working towards prediction (Leary, 1984b, 1985a). His complex interactive geometric models have proven useful in several different scientific disciplines (entomology (Mattson and Addy, 1975), forestry (Leary, 1972, 1975, 1985b), ecology (Odum, 1971, Dindal, 1975), education (Leary, 1984a), and psychology (Lodge, 1972)). They have also proven useful to scientists whose focus is synthesis and whose efforts are searchable on the internet at the websites of Anthony Judge and Timothy Wilken. The Nobel laureate Albert Szent-Györgyi is known for having said, 'Research is to see what everybody else has seen, and to think what nobody else has thought'. One could argue that Ed Haskell's work was incomplete and at risk of overreaching, as a comprehensive predictive model for unified science, in light of Quine's analyses. On the other hand one could argue that Quine's prescription that synthesis occur at the level of explanation may not be realistic, given the diversities of nature and methods of studying it.

It should be noted that the present analysis is also incomplete and that this topic presents a fertile field for further research. More extensive correspondence and unpublished papers are available beyond those already archived at Harvard. Further information may be obtained by contacting the authors at the email addresses listed in the Notes on Contributors.

* **Cited Letters** (in chronological order by writer from sources as shown)

>AL personal files: Ann Lodge, Santa Fe, NM
>DBQ personal files: Douglas Quine, Bethel, CT
>RAL personal files: Rolfe Leary, St. Paul, MN
>WVQ Archives: Harvard University Houghton Library, Cambridge, MA

- EFH to WVQ, December 24, 1945; quote from letter in WVQ Archives
- EFH to Langlois, June 12, 1948; quote from letter in RAL files
- EFH to WVQ, June 28, 1955; quote from letter in WVQ Archives
- EFH to WVQ, June 30, 1959; quote from letter in RAL files
- EFH to WVQ, October 12, 1972; quote from letter in RAL files
- EFH to WVQ, October 14, 1983; quote from letter in RAL files
- HGC to EFH and WVQ, January 18, 1946; quote from letter in AL files
- HGC to WVQ, April 4, 1954; quote from letter in AL files
- HGC to Mesthene, September 2, 1965; quote from letter in WVQ Archives
- HGC to WVQ, June 11, 1989; quote from letter in DBQ files
- WVQ to EFH, June 24, 1939; quote from letter in WVQ Archives
- WVQ to EFH, July 22, 1939; quote from letter in WVQ Archives
- WVQ to EFH, January 4, 1946; quote from letter in WVQ Archives
- WVQ to HGC, April 9, 1954; quote from letter in AL files
- WVQ to EFH, June 20, 1955; quote from letter in WVQ Archives
- WVQ to EFH, February 13, 1967; quote from letter in WVQ Archives
- WVQ to EFH, September 22, 1969; quote from letter in WVQ Archives

- WVQ to HGC, October 1, 1989; quote from letter in DBQ files

Notes

1. This reminds co-author RAL of the Tee shirt he has wanted to market: 'When the data don't fit my model, like Mendeleev, I change them!'.
2. Haskell's personal diary states: 'Men's Class Paper: "Control of Power by Values". Big success: Thomas Finletter [*who became NATO Ambassador for the Kennedy administration in 1961*] was most helpful. De Bessenyey at least fair'.

References

Bohr, N. (1955) 'Analysis and Synthesis in Science'. In O. Neurath, C. Morris, & R. Carnap (eds.) 1955, *International Encyclopedia of Unified Science*, vol. 1, Part 1. Chicago, IL: University of Chicago Press, p. 28.

Cassidy, H.G. (~1991) *Edward Froelich Haskell's Legacy: Interaction Geometry and A Classification of the Sciences*, unpublished manuscript.

Dindal, D. (1975) 'Symbiosis: Nomenclature and proposed classification', *The Biologist*, 57(4): 129–142.

Einstein, A., & Infeld, L. (1938) *The Evolution of Physics* (New York, NY: Simon and Schuster).

Haskell, E.F. (1940) 'Mathematical Systematization of "Environment", "Organism", and "Habitat"', *Ecology*, 21: 1–16.

Haskell, E.F. (1947) 'A Natural Classification of Societies'. In Roy Waldo Miner (ed.) *Transactions of the New York Academy of Sciences, Series II*, vol. 9(3). New York, NY: The New York Academy of Sciences, pp. 186–196.

Haskell, E.F. (1949) 'A Clarification of Social Sciences'. *Main Currents in Modern Thought*, vol. 7. New Rochelle, NY: Center for Integrative Education, pp. 45–51.

Haskell, E.F. (1960) *The Control of Power by Values* unpublished manuscript.

Haskell, E., Cassidy, H.G., Clark, J.W., & Jensen, A.R. (1972) *Full Circle: The Moral Force of Unified Science. Current Topics of Contemporary Thought*, vol. 8. New York, N.Y.: Gordon and Breach.

Haskell, E.F., & Cassidy, H.G. (~1976) *Evolution and Education: Their System-Hierarchic Structure and Ethical Orientation*, unpublished manuscript.

Haskell, E.F. (1980) 'Expanding and Contracting Simultaneously', *Oberlin Alumni Magazine*, 76(2): 35–41.

Haskell, E.F., & Cassidy, H.G. (ed.) (~1990) *Assembly of the Sciences into a Single System: Unified-Science*. (unpublished Word manuscript in RAL files).

Judge, Anthony. http://www.laetusinpraesens.org/docs00s/cardioid.php

Langlois, T.H. (1936) *A Study of Small-Mouth Bass Micropterus Dolomieu (Lacepede) in Rearing Ponds in Ohio* (Columbus, OH: Ohio Biological Survey).

Leary, B.B. (1984a) 'What Place Are You In?', *Academic Therapy*, 19(3): 269–275.

Leary, R.A. (1972) 'Estimation of coaction from experimental data', Paper delivered at *The First International Conference on Unified Science*, Waldorf Astoria Hotel, November 23–26, New York, NY.

Leary, R.A. (1976) 'Interaction geometry: An ecological perspective' USDA Forest Service GTR-NC-22.

Leary, R.A. (1984b) 'New Directions for Scientist's Research', Paper delivered to the Faculte de Foresterie et Geodesie, Laval University, Quebec, Canada, January 18, 1984.

Leary, R.A. (1985a) 'A Framework for Assessing and Rewarding a Scientist's Research Productivity', *Scientometrics*, 7(1–2): 29–38.

Leary, R.A. (1985b) *Interaction Theory in Forest Ecology and Management* (Dordrecht: Martinus Nijhoff/ Dr. W. Junk Publishers).

Lodge, G.T. (1972) 'Cooperation and Conflict within the Personality', *Abstract Guide of XXth International Congress of Psychology*, Tokyo, p. 583.

Mattson, W.J. and N.D. Addy. (1975) 'Phytophagus Insects as Regulators of Primary Production in Forest Ecosystems', *Science*, 190(4214): 515–522.

Neurath, O. (1955) 'Unified Science as Encyclopedic Integration', In O. Neurath, C. Morris, R. Carnap (eds.) 1955, *International Encyclopedia of Unified Science*, vol. 1, Part 1. Chicago, IL: University of Chicago Press, pp. 1–27.

Odum, E.P. (1971) *Fundamentals of Ecology*, third edition (Philadelphia, PA: W. B. Saunders Company).

Quine, W.V. (1972) 'Levels of Abstraction', Paper delivered at *The First International Conference on Unified Science*, Waldorf Astoria Hotel, November 23–26, New York, NY.

Quine. W.V. (1985) *The Time of My Life: An Autobiography* (Cambridge, Mass: The MIT Press).

Schwemmler, W. (1984) *Reconstruction of Cell Evolution: A Periodic System* (Boca Raton, FL: CRC Press).

Wilken, Timothy. http://www.synearth.net/KU1/UCS-Basics3.html

Part III

Quine's Connection with Pragmatism

7

The Web and the Tree: Quine and James on the Growth of Knowledge

Yemima Ben-Menahem

Quine's metaphor – the web of belief that touches on experience at its periphery, but gets exceedingly removed from experience as one approaches its center – is widely known to philosophers. 'Two Dogmas of Empiricism', the classic paper in which the metaphor first appeared, and the various theses that Quine sought to illustrate by means of his metaphor have all received a great deal of attention. Far less attention has been paid to the strikingly similar features of James's epistemology; indeed, the similarity between the two is mostly overlooked. One reason for this oblivion to the similarity between Quine and James might be related to a prevalent *mis*reading of James, on which his pragmatic theory of truth identifies truth with utility and sanctions wishful thinking. Such abuse of rationality (this misreading suggests) can hardly count as epistemology, let alone be compared with that of Quine. Another reason may have to do with philosophical context: Quine interacted more directly with the logical positivists than with the pragmatists. He engaged in continuous dialogue with Carnap and often mentioned Neurath, but did not devote comparable efforts to critical dialogue with the pragmatist tradition. James, in particular, is almost completely absent from his writings. The comparison undertaken here may not only shed some light on the ancestry of Quine's epistemology, but can encourage replacement of the common stereotype of James with a more balanced interpretation. I do not intend to stretch the analogy beyond its limits – obviously Quine and James differ on significant philosophical issues – nor do I wish to underrate Quine's affinity to other pragmatist philosophers such Peirce and Dewey. I believe, however, that the centrality of Quine's metaphor in his own writings as well as Twentieth-Century epistemology at large warrants a close look at its precursors. Beginning with the analogies between the epistemic metaphors of James and Quine, this paper proceeds to examine their views on a number of other characteristic positions of pragmatism such as the rejection of foundationalism and skepticism and the acknowledgement of the social dimension of knowledge. It then comments on some of the differences between Quine and James and concludes with an examination of Quine's own assessment of his relation to the pragmatist tradition.

1   Quine's web

Let us take a look at Quine's familiar metaphor:

> The totality of our so-called knowledge or beliefs, from the most casual matters of geography and history to the profoundest laws of atomic physics or even of pure mathematics and logic, is a man-made fabric which impinges on experience only along the edges. Or, to change the figure, total science is like a field of force whose boundary conditions are experience. (1953:42)

This image illustrates for Quine a number of features central to his epistemology:

* Empirical import: Typically, problems arise when the existing web (or part of it) runs into conflict with some new experience. Such problems then move inwards from the periphery to the interior as we try to adjust the web to the new situation. Experience may thus impact any segment of the web and harmony with experience constitutes the major epistemic constraint on web construction.
* Holism: 'Our statements about the external world face the tribunal of sense experience not individually, but only as a corporate body' (1953:41). 'No particular experiences are linked with any particular statements in the interior of the field, except indirectly through considerations of equilibrium affecting the field as a whole' (1953:42). And again: The connections between surface irritations constitute 'a maze of intervening theory' (1960:275). Still later, the web is described as consisting of 'ninety nine parts conceptualization to one part observation' (1981b:97). In retrospect (1991:268), Quine qualifies this holism, allowing parts of the web, rather than only the web as a whole to be confirmed. He also tells us (1991:269) that Pierre Duhem's holism was brought to his attention only after delivering the 'Two Dogmas' talk in 1950, upon which he added an acknowledging footnote (no.17) to the published paper.

* No analytic-synthetic dichotomy: There are no privileged sentences, analytic, *a priori* or necessary, no sentences that are completely detached from the experimental boundary, none that can be affirmed merely on the basis of their own indubitable force. Even logical laws are 'simply further statements of the system' (1953:42), merely more central and perhaps more stable than experiential-peripheral ones due to the multitude of inferences they participate in. For Quine, it is this point which makes his position 'a more thorough pragmatism' than that of 'Carnap, Lewis, and others' (1953:46). It is noteworthy that Quine mentions Carnap and Lewis as typical proponents of pragmatism rather than the founding fathers of that tradition. Since the analytic was construed by the logical positivist as conventional, Quine also expresses his rejection of the analytic-synthetic distinction in terms of the blend of fact and convention. The famous concluding passage of 'Carnap and Logical Truth' says: 'The lore of our fathers... is a pale gray lore, black with fact and white with convention. But I have found no substantial reasons for concluding that there are any quite black threads in it, or any white ones' (1966[1954]:125).
* Underdetermination: 'The total field is so underdetermined by its boundary conditions, experience, that there is much latitude of choice as to what statements to reevaluate in the light of any single contrary experience' (1953:42). Note the difference between two situations of underdetermination that are often conflated. In 'Two Dogmas', Quine first asserts that 'any statement can be held true come what may, if we make drastic enough changes elsewhere in the system' (1953:43). In the situation envisaged in this quote, theories that have conflicting empirical implications (i.e., are *not* empirically equivalent), may still retain the same hypothesis h. In other words, h could be saved (in a different framework) despite the fact that the original theory that contained it had been refuted. But Quine also considers another case of underdetermination, which occurs when there are empirically equivalent theories, namely theories that have exactly the same empirical implications and yet contain incompatible theoretical statements. Underdetermination in both of these cases depends on holism. In later years it is the existence of empirically equivalent theories that Quine usually means by underdetermination. However, when he realized that empirically equivalent theories that appear to be incompatible may in fact be merely alternative formulations of the same theory, he began to question underdetermination of this kind.[1]
* The fabric is man-made. We do not simply find or discover truth, according to Quine. First, since only sentences can be true or false, we

must create language to formulate them. Languages and their categories are not forced upon us by experience but are human creations that involve a great deal of latitude. Second, even when in possession of a language, we cannot use it to deduce laws from experience – this is the notorious problem of induction. Rather, we must do with the reverse derivation and infer observation sentences from more general, theoretical ones. These inferences, 'observation categoricals', as Quine calls them, serve to test the consistency of the web with experience. Quine reaffirmed the man-made character of the web in later writings. 'Despite my naturalism, I am bound to recognize that the systematic structure of scientific theory is man-made. It is made to fit the data, yes, but invented rather than discovered, because it is not uniquely determined by the data' (1981a:33).

* Pragmatic criteria: Quine refers to the criteria that guide us in the construction and modification of the web as pragmatic, but does not spell out the precise character of this pragmatism. He seems to use the term in its daily sense, not in any technical sense that would link it with the pragmatist philosophers. The need for further criteria arises from underdetermination: mere fit with experience will not suffice to determine the structure of the web uniquely. Given our scientific heritage on the one hand and 'a continuing barrage of sensory stimulation' on the other, we must use further criteria to choose from the empirically equivalent options we have. These criteria, Quine says 'are, where rational, pragmatic' (1953:46). It is important that Quine sees pragmatic criteria as rational; in contrast, presumably, with other criteria such as authority and superstition which may also happen to play a role in shaping our belief-system, but are not considered rational. Quine later condensed the rational considerations into a single guideline – *the maxim of minimum mutilation*—we make the changes we take to be necessary while keeping as much as we can of the existing structure.

2   James's tree

Here is one of the images James uses in his *Pragmatism* to describe our belief system. The body of truth 'grows much as a tree grows by the activity of a new layer of cambium' (1955 [1907]:52). Less metaphorically, he asserts that our system grows by 'interpreting the unobserved by the observed' (1955 [1907]:55). It is instructive to go over the above Quinean list and take note of its parallels in James.

* Empirical Import: James makes it repeatedly clear that the principal epistemic desideratum is conformity with experience. James does not use Quine's terms of surface irritations, nerve endings, and so on. In fact, he sometimes explicitly endorses a notion of experience broader than that of Quine (*e.g* in1955 [1907]:44). Nonetheless, James's empiricist conception of knowledge is evident not only in *Pragmatism*, but also in *The Principles of Psychology* (1890), his pioneering attempt to turn psychology into an empirical science. James's portrayal of Pragmatism as 'a new name for some old ways of thinking' (in the subtitle to the Lectures on Pragmatism) further indicates his perception of pragmatism as continuous with empiricism. On the issue of empiricism versus rationalism, the former adopted by those whom James refers to as the 'tough-minded', the latter by the 'tender-minded', James certainly sides with the tough-minded empiricists (Lecture I). The following quotes indicate the prevalence of James's demand for a solid basis in experience. The distance between this firm commitment to empirical support and the common image of James as sanctioning irresponsible make-believe should be obvious.

Truth lives, in fact, for the most part on a credit system. Our thoughts and beliefs 'pass,' so long as nothing challenges them, just as banknotes pass as long as nobody refuses them. But this all points to direct verifications somewhere, without which the fabric of truth collapses, like a financial system with no cash-basis whatever. ... Beliefs verified concretely by somebody are the posts of the whole superstructure. (1955 [1907]:52)

For pluralistic pragmatism, truth grows inside of all finite experiences. They lean on each other, but the whole of them, if such a whole there be, leans on nothing. All 'homes' are in finite experiences; finite experience as such is homeless. Nothing outside of the flux secures the issue of it. It can hope salvation only from its own promises and potencies. (1955 [1909]:169)

Our theory must mediate between all previous truths and certain new experiences. It must derange common sense and previous beliefs as little as possible, and it must lead to some sensible terminus or other that can be verified exactly. (1955 [1909]:142)

But all roads lead to Rome, and in the end and eventually, all true processes must lead to the face of directly verifying sensible experiences *somewhere*. (1955 [1909]:141)

The word 'somewhere' is important, for James acknowledges that verification is not of individual sentences or ideas but of inter-related systems. It is a prolonged, long-term process to which many individuals contribute.

* Holism: James is more concerned to critique foundationalism than reductionism. These themes are closely related, of course, since the reduction required by the reductionist is always a reduction to a preferred foundation (more on this below). Arguing against reductionism, Quine emphasizes that, in general, individual sentences do not come with their own observable implications and can therefore be tested only in conjunction with larger chunks of the web. Arguing against foundationalism, James emphasizes that we do not face experience empty-handed and try to make sense of it individually, with no background of previous knowledge. Rather, we use the elaborate system we have inherited to interpret our own experiences and change the system only when no way of keeping it intact is available. Although the term 'holism' is not used by James, the picture that emerges is just as holistic as that of Quine. Both of them hold that we aim at equilibrium in our system in its entirety. Holism is also manifest in their respective accounts of the dual traffic between theory and experience. Whereas we typically change theory to accommodate recalcitrant experiences, both Quine and James also countenance the reverse process, whereby we sacrifice an observation sentence (or reinterpret it) in order to save parts of our theory. The feasibility of this option speaks against a simplistic picture of observation as a fixed basis to which every theoretical sentence can be reduced.

The novelty soaks in; it stains the ancient mass; but it is also tinged by what absorbs it.... It happens relatively seldom that the new fact is added *raw*. More usually it is embedded cooked, as one might say, or stewed down in the sauce of the old. (1955 [1907]:83, Italics in the original)

* No analytic-synthetic dichotomy: Here James makes another use of the tree-image. '*Truth also has its paleontology*'. What traditional (rationalistically-minded) philosophers see as eternal and incorrigible truths are for James only '*the dead heart of the Living tree*' (1955 [1907]:53). Not completely dead, however, for 'how plastic even the oldest truths nevertheless really are has been vividly shown in our day by the transformation of logical and mathematical ideas, a transformation which seems even to be invading physics' (1955 [1907]:53). James contrasts the traditional conception of necessary truths with

the pragmatic conception he is arguing for. The theorems of logic and mathematics and even some of the laws of nature were once conceived as representing 'the eternal thoughts of the Almighty. His mind also thundered and reverberated in syllogisms. He also thought in conic sections, squares and roots and ratios, and geometrized like Euclid' (1955 [1907]:48). In fact, however (James argues), all of these laws 'are only a man-made language, a conceptual short-hand ... in which we write our reports of nature; and languages, as is well known, tolerate much choice of expression and many dialects' (1955 [1907]:48–49).

* Underdetermination: James is aware of the possibility of underde-termination and draws the conclusion that when two theories are equally well supported by experience, criteria other than empirical support come into play.

Yet sometimes alternative theoretic formulas are equally compatible with all the truths we know, and then we choose between them for subjective reasons ... taste included, but consistency both with previous truth and with novel fact is always the most imperious claimant. (1955 [1909]:142)

* The system is man-made: James could not have been more explicit: 'The trail of the human serpent is thus over everything' (1955 [1907]:53). James (as we have seen in the quotes on necessary truth) regards the system as human for the very same reason adduced by Quine, namely, that language, with its categories and classifications, is a human crea-tion. Experience does not come ready-made with its 'proper' descrip-tion; it does not wear names, predicates and relations, on its sleeve. There is no privileged language that can be singled out as a true description of reality, no ideal language that nature should have used to describe itself, so to speak.

* Pragmatic criteria: As for methodological guidelines, the reasonable method, according to James, aims at 'a minimum of disturbance' to the existing system. The idea is not only identical with that of Quine in terms of substance; it also uses the same terminology.

[We] preserve the older stock of truths with a minimum of modification ... A *outrée* explanation, violating all of our preconceptions, would never pass for a new account of a novelty. ... [New truth] marries old opinion to new fact so as ever to show a minimum of jolt, a maximum of continuity. We hold a theory true just in proportion

to its success in solving this "problem of maxima and minima".
(1955 [1907]:50–51)

To sum up: James and Quine share important features of their epistemology: They propose a man-made interconnected system which is
responsible to experience and yet underdetermined by it. The system
is seamless in the sense that it has no privileged truths, dynamic in
allowing variation of each one of its components, and economic when
obeying the maxim of minimum mutilation.

As far as I know, Quine nowhere mentions the similarity between his
ideas and those of James. According to his scientific autobiography in
the *Library of Living Philosophers* volume, however, James's *Pragmatism*
was one of the only two philosophy books Quine read as a teenager. 'I
read them compulsively and believed and forgot all' (1986:6).

3   Other characteristics of pragmatism

Pragmatism, like other philosophical schools, cannot be given a precise
definition, but there are a number of characteristics that recur in pragmatists' writings. As we will see, most of these are also shared by Quine.
Let me briefly sketch the traits that I see as typical of pragmatism. I put
most of these in negative terms, i.e. in terms of traditional positions that
pragmatists object to. These objections provide a broader context for
the above-mentioned positive theses. (I set aside the ethical and polit-
ical dimensions of pragmatism, which are of little relevance to Quine's
philosophy.)

No Foundationalism: A major difference between Seventeenth-Century
epistemology and that of the American pragmatists pertains to foundationalism. In the Seventeenth Century, erasing all previous belief and
making a fresh start was considered to be the right epistemic method.
The underlying metaphor was that of a building – a firm foundation
and a systematic construction ensures its stability. Both discovery and
justification were thought to be taken care of by this procedure. Peirce,
however, was strongly opposed to the Seventeenth Century-recipe.
Whatever the firm foundation was supposed to be, whether it consisted
of self-evident truths or bare sense data, the foundationalist method
could not work. Peirce's critique of Descartes on the one hand, and the
empiricists on the other, was that one can neither destroy the entire body
of previous belief, nor construct a new one from scratch. Instead, investigation always begins with a localized problem, an island of doubt in a
sea of beliefs that one does not question. We have seen above that James

also stresses our dependence on a vast amount of traditional knowledge. A total revision of the system cannot even be conceived. Quine agrees. Although his web has a boundary in experience, this boundary does not function as a foundation in the Seventeenth-Century sense. The inner parts of the web are neither reducible to experience nor deducible from it and are not constructed serially in any one direction. Moreover, the rejection of foundationalism is connected with several other canons of Quine's philosophy: His denial of prior philosophy and affirmation of the continuity of philosophy with the sciences, his immanent conception of truth (see below) and his rejection of what he succinctly calls 'cosmic exile'.

> The philosopher's task differs from others', then, in detail; but in no such drastic way as those suppose who imagine for the philosopher a vantage point outside the conceptual scheme that he takes in charge. There is no such cosmic exile. (1960:275)

No Skepticism: Skepticism, even if only a methodological point of departure as in Descartes, and certainly when adopted as a sustained epistemic position, is an anathema to pragmatists. The space drained of belief, they claim, the space in which skepticists presume to be living in a blissful state of suspended judgment, is unfit for human beings. Skepticists may be right to hold that perfect justification is impossible, but, pragmatists retort, so is global doubt. Even local doubt cannot be triggered at will, out of blind obedience to a methodological instruction. It must be a 'living doubt'.

> Some philosophers have imagined that to start an inquiry it was only necessary to utter a question or set it down on paper, and have even recommended us to begin our studies with questioning everything! But the mere putting a proposition in the interrogative form does not stimulate the mind to any struggle after belief. There must be a real and living doubt, and without all this, discussion is idle. (Peirce, 1935 [1877]: section IV)

Moreover, it is often assumed that belief must be justified whereas doubt needs no reasons. By contrast, pragmatists maintain that doubt too requires justification. James argued further that in some cases belief beyond evidence is justified – his notorious 'will to believe' (see below). The rejection of skepticism and foundationalism go hand in hand in recognizing that we can only address localized problems set against a background of what is taken for granted. Even though this background is neither intrinsically beyond doubt, nor perpetually stable, it is 'good enough' in guiding us towards further knowledge. The provisional status of background beliefs allows pragmatists to reject skepticism while, at the same time, endorsing falliblilism.

Fallibilism: Pragmatists repeatedly stressed that there are no incorrigible beliefs. While we cannot revise the entire system of knowledge at once, we must often revise some of its components. (Recall, however, that revision of one part may eventually affect others.) For both James and Quine revisionism applies to any fraction of the system, from observation reports to logical and mathematical truths. Here is how Quine combines all of these pragmatist theses, linking them with holism:

> The naturalist philosopher begins his reasoning within the inherited world theory as a going concern. He tentatively believes all of it, but believes also that some undistinguished portions are wrong. He tries to improve, clarify, and understand the system from within. He is the busy sailor adrift on Neurath's boat. (1981:72; 1981a:33)

No Essentialism: The pragmatist theory of meaning, in both its Peircean and Jamesian versions, stands in marked contrast with essentialism. Indeed, it was the aim of these thinkers to develop an alternative to the traditional conception on which meanings consist in fixed essences that our definitions then strive to capture. Their alternative was the dynamic and empiricist approach to meaning that has become the emblem of pragmatism. Blaming essentialism for a series of philosophical blunders, James finally condemns it as no less than a form of magic.

> Metaphysics has usually followed a very primitive kind of quest. You know how men have always hankered after unlawful magic, and you know what a great part, in magic, *words* have always played. If you have his name, or the formula of incantation that binds him, you can control the spirit, genie, afrite, or whatever the power may be.... The universe has always appeared to the natural man as a kind of enigma, of which the key must be sought in the shape of some illumination or power-bringing word or name. That word names the universe's principle, and to possess it is, after a fashion, to possess the universe itself. "God", "Matter", "Reason", "the Absolute", "Energy", are so many solving names. You can rest when you have them. You are at the end of your metaphysical quest. (1955 [1907]: 46 italics in original)

Quine's work on meaning is much more elaborate than that of James's and critique of essentialism is only its starting point. Quine moves on to critique what he terms the 'myth of the museum' according to which words correspond with either distinct mental ideas or distinct external references (1969:27). His holistic conception of meaning, which, he argues, involves the indeterminacy of translation and the relativity of reference, is as far removed from essentialism as one can get.

The Social Dimension of Knowledge: In the Seventeenth Century, epistemology was generally conceived in terms of the mental activity of an individual. Pragmatists, on the other hand, stress the social character of language and consequently also of epistemology. Recall Peirce's theory of signs. In addition to the sign and the signified, there is always an interpreter who can be an immediate addressee as well as a future one. Meaning is generated in the prolonged interaction between speakers and interpreters. Likewise, the creation of knowledge is a long-term social process carried out by a community of investigators. We have seen that the need for a linguistic and epistemic tradition is also acknowledged by James. The social dimension of language is manifest in *Word and Object*, where Quine characterizes even observation sentences in social rather than individualistic terms. They are occasion sentences 'on which there is to be firm agreement on the part of well-placed observers' (1960:44). Note that this way of characterizing observation sentences is very different from its usual empiricist characterization in terms of individuals' sense impressions, for it allows sentences that we do not usually consider observational, such as 'God is angry' to pass for observation sentences. If, for instance, the community in question accepted a rule to the effect that whenever it rains, and only then, God is angry, then its members would assent to 'God is angry' when it rains, and dissent when it doesn't. According to Quine's criterion, in this community 'God is angry' is an observation sentence!

Belief and Action: Pragmatists maintain that belief manifests itself in action. Inspired by Alexander Bain's 1859 *The Emotions and the Will*, this conception is appealing to pragmatists in its focus on concrete observable consequences rather than obscure mental states. Quine does not make much of the connection between belief and action but, seeing it as an aspect of empiricism, he approve of it (1981a).

No Correspondence Theory of Truth: James ridicules the notion of an 'absolute correspondence of our thoughts with an equally absolute reality' (1955 [1907]:54). The world is given to us in language, and language, as we have seen, is conceived by both James and Quine as a human creation. We compare our various descriptions of the world with one another, but we cannot compare them with bare reality, pre-linguistic fact, a thing in itself. Peirce and Dewey have also argued against the correspondence theory, and against the naïve realism it gives rise to. Despite their agreement on the flaws of the correspondence theory, however, pragmatists vary significantly on the positive accounts they offer in its stead. 'The pragmatic theory of truth' is therefore a misleading term. Peirce defined truth as 'the opinion which is fated to be ultimately agreed to by all who investigate'. And he took the term 'fate' seriously, though not in the sense of reifying fate or seeing it as produced by a conscious being: 'This activity of thought by which we are carried, not where we wish, but to a foreordained goal, is like the operation of destiny' (1935 [1878]: section IV). Neither James nor Quine were willing to follow Peirce in this teleology of truth. Quine notes (1981a:31) that we have no way of comparing theories in terms of their similarity, a comparison that is presupposed by Peirce's definition. In contrast to Peirce, James had a pluralistic conception of truth. He distinguished between different kinds of truths: verifiable and unverifiable, well-rooted and novel, those on which we can postpone our verdict, and those that must be urgently decided on, those that are independent of our belief in them and those that are up to us to *make* true. Quine was just as opposed to James's ideas about truth as he was to Peirce's. He characterizes his own concept of truth as immanent: 'we are always talking within our going system when we attribute truth; we cannot talk otherwise' (1981a:34). It is our best scientific theory that tells us what is true and what is real. To accept this theory and still refuse to acknowledge its truth (or the reality of the entities it invokes), is senseless, according to Quine. At the same time, to uphold a notion of truth firmer and more fundamental than that tied to the best theory we have is just as senseless. The immanence of truth goes hand in hand with 'unregenerate realism, the robust state of mind of the natural scientist who has never felt any qualms beyond the negotiable uncertainties internal to science' (1981a:28). Consequently, for Quine, 'physical objects are real, right down to the most hypothetical of particles, though this recognition of them is subject, like all science, to correction' (1981a:33). He therefore differs from both Peirce and James on the subject of truth. The differences between James and Quine are examined in more detail in the next section.

4   On some differences between James and Quine

Despite the similarity between their epistemic models, Quine and James were rather different in their philosophical temperaments. As much as

James cherished the scientific method, science was for him only one facet of human experience. Art, religion, the moral life and even metaphysics had to find their place, alongside science, in his world view. Quine did not share these concerns. Whereas Quine (his critique of logical positivism notwithstanding), was mainly interested in meaning in the narrow sense of the term, James aspired to create a space for meaning in the broader sense associated with concerns over the meaningful life. I would venture the claim (without arguing it here) that, ultimately, it was the moral dimension of life, rather than the scientific dimension, that made life meaningful for James. Hence, in particular, the importance he ascribed to the problem of free will. His pluralism was intended to enable peaceful co-existence between the different aspects of life. Such co-existence is possible, according to James because different notions of truth and different methods of ascertaining truth are applicable to distinct types of problems so that they need not run into conflict with one another. Thus, ordinary questions about facts must be decided by direct empirical evidence. Other questions, requiring a more thorough investigation and resisting an immediate answer could also be eventually settled by science. But there are questions for which there is no hope of getting a scientific resolution, and with regard to these, other considerations come into play. For example, citing Poincare, James contends that some mathematical issues can be decided by fiat.[2] His most famous plea for deviation from evidence-based belief involves a particular kind of existential decision that James considered to be impervious to science. It is with regard to such decisions that he allowed the leap of faith that gave him such bad reputation.

James puts forward several conditions that make a statement a legitimate candidate for 'the will to believe', or 'the right to believe', the term he would later prefer. First, as mentioned, the statement cannot be confirmed (or refuted) by science. James maintained, for example, that the problem of free will cannot be decided on the basis of empirical evidence. Second, the alternatives in question must be 'live options' for the person who deliberates. Buddhism may be a live option for people who do not entertain Christianity, and so on. James requires that both of the alternatives, accepting or denying Buddhism, say, be live options for the deliberating person. Third, the alternatives must be such that a decision is forced; no neutral state of suspended judgment is available. When one considers accepting a job or pursuing a relationship, a negative decision is just as consequential as a positive one. Fourth, the issue in question should be momentous rather than trivial. The most important condition in my view, however, is the self-fulfilling nature of the belief – *'where faith in fact can help create the fact'*. When this condition is satisfied, James continues, only 'an insane logic' would preclude 'faith running ahead of scientific evidence' (1956 [1897]:25, italics in the original). James draws a famous analogy: 'We stand on a mountain pass in the midst of whirling snow and blinding mist' (1956 [1987]:31). Shall we stay on the mountain – and freeze to death – or try to make our way down? The alternatives are indeed living, forced and momentous. And our chance to succeed may be enhanced by confidence. Similarly, confidence in friendship or recovery may help bring about friendship and recovery. Optimism beyond existing evidence is thus defensible. (Note that in such cases the optimistic belief might still be refuted in the future.) However, James's primary example – religious faith – remains controversial. While constituting a living, forced and momentous option, it does not seem to be self-fulfilling.

Even if we set aside the very special cases in which James allowed a leap of faith, his conception of meaning in general, the conception presented in *Pragmatism* (rather than that of the earlier *The Will to Believe*) is also different from that of the empiricist. For the empiricist only empirically testable implications confer meaning, while for James a "difference that makes a difference" can be a difference to the life of the *believer*. A statement that cannot be tested and is therefore meaningless by empiricist standards – an assertion of providence, say – can still affect the life of the believer and would thus be meaningful for James. Quine could not accept such implications for the believer as empirical implications.[3]

Severe critique by leading philosophers such as Russell (1966[1910]) and Moore (1922[1907]) nearly ruined James's stature as a responsible philosopher. Although James repeatedly stressed the connection between pragmatism and empiricism, the frivolous image persisted. He has not only been misinterpreted but also derided as incapable of coherent thought and perhaps not even aiming at it. 'Clarity and consistency were not James family traits. Part of the problem is that James's philosophically grew up in the later Nineteenth Century, an era in which ambiguity, indirection and rococo encrustations of metaphor were standard features of philosophical expression' (Kirkham 1992:77–78).[4] The first section of this paper should have convinced the reader that there is much more to James's epistemology than the contentious will to believe. Moreover, it is the particular application to belief in God that is troublesome; other applications of James's license to believe, such as confidence in oneself or belief in friendship may be innocent. Nonetheless, it is the problematic example of religious faith that tainted James's philosophy in the eyes of many of his readers, including Quine, who, when finally coming to evaluate his position vis-a-vis the pragmatist tradition, ignored the parallels between James's philosophy and his own, referring only to 'James's notorious defense of wishful thinking' (1981a: 32).

5 Quine's reflections on his relationship to the pragmatist tradition

In 1975, Quine was invited to deliver a paper at a conference on 'The Sources and Prospects of Pragmatism'. His paper appeared in 1981 in two versions: the full version, entitled 'The Pragmatists' Place in Empiricism', in a volume containing the conference papers (Quine, 1981a), and an abridged version, under the name 'Five Milestones of Empiricism' in *Theories and Things* (Quine, 1981). The latter contains only part of the former, excluding any discussion of pragmatism. The fact that Quine chose to include in his collection only the sections on the five milestones of empiricism indicates more than anything he said explicitly that he ascribed little significance to the impact of pragmatism on his own thought, and was perhaps also doubtful about the importance of pragmatism in general. The reader gets the same impression from what Quine does say about pragmatism, albeit graciously, in the unabridged version of the paper. His main complaint is that 'it is not clear...what it takes to be a pragmatist....the term "pragmatism" is one we could do without' (1981a:23). What pragmatists share, according to Quine, is empiricism, even if not the specific brand of empiricism he commends. Hence, the 'five points where empiricism has taken a turn for the better' (1981a:23), constituting the milestones on the road to Quine's own empiricism:

1 The shift from ideas to words.
2 The shift from terms to sentences.
3 Holism – the shift from sentences to systems of sentences.
4 No analytic-synthetic dualism.
5 Naturalism – no prior philosophy.

Quine ascribes the first of these transitions to John Horne Tooke's critique of Locke and the second to Jeremy Bentham. The founding fathers of analytic philosophy, Frege, Russell and Wittgenstein as well as the logical positivists are also mentioned by Quine as promoting these two insights. The remaining three transitions are characteristically Quinean even if he deemphasizes his role as their proponent. For example, he cites August

Comte as the originator of naturalism. Notably, none of the five points is attributed by Quine to the pragmatists. In the ensuing discussion of the position of pragmatists on his five points, he mentions various disagreements with them. He criticizes Peirce for vacillating between words and ideas and between beliefs and sentences (although ultimately settling for sentences) and for not being sufficiently outspoken about holism. He criticizes James for being kind to wishful thinkers and both James and Dewey for declining (immanent) realism. He disagrees with Lewis about the analytic-synthetic distinction. Quine mentions in passing a couple of points of agreement with the pragmatists, fallibilism, the repudiation of Cartesian doubt and the recognition of Darwinism as a basis for understanding of the human mind and its conceptual categories. There are, however, two points for which Quine credits the pragmatists: the man-made nature of truth and the social character of meaning. To the latter he refers as 'behavioristic semantics', stressing that it was Dewey, rather than Wittgenstein, who first insisted 'that there is no more to meaning than is to be found in the social use of linguistic forms' (1981a:36–37). The concluding lines of Quine's generally critical paper on pragmatism are more generous than its opening ones: Although he repeats the complaint that he 'found little in the way of shared and distinctive tenets' he goes on to say: 'The two best guesses seemed to be behavioristic semantics, which I so heartily approve, and the doctrine of man as truth-maker, which I share in large measure' (1981a:37).

Despite this acknowledgement, it remains a fact that Quine saw none of the five advances in empiricism as initiated by pragmatism, and that he omitted the entire discussion of pragmatism from the version included in his collection. The argument of the present paper was that Quine had more in common with pragmatism in general and James's epistemology in particular than his reflections disclose.

Notes

1. See Quine (1975). On the difficulties in illustrating and demonstrating empirical equivalence and on Quine's change of mind with regard to this thesis, see chapter 6 of my 2006. A recent proof of empirical equivalence can be found in Putnam (2012). I am unaware of a general proof of the other type of underdetermination.
2. James (1955) [1907]:49, p. 237. See also his (1956):15.
3. Quine finds an ambiguity between the two kinds of empirical implication already in Peirce. He also notes that James admitted this ambiguity in a letter to Lovejoy (Quine, 1981a:33).

4. James's colorful language is perhaps partly to blame. For example, when contrasting the pragmatist with the rationalist, whom he describes as 'of a doctrinaire and authoritative complexion', he says: 'A radical pragmatist on the other hand is a happy-go-lucky anarchistic sort of creature' (1955 [1909]:168).

References

Bain, A. (1859) *The Emotions and the Will* (London: J.W. Parker and Sons).

Ben-Menahem, Y. (2006) *Conventionalism* (Cambridge: Cambridge University Press).

James, W. (1955 [1907][1909]) *Pragmatism and four essays from The Meaning of Truth* (New York: Meridian Books).

James, W. (1956 [1897]) *The Will to Believe and Other Essays in Popular Philosophy* (New York: Dover).

Kirkham, R.L. (1992) *Theories of Truth* (Cambridge Mass : Bradford MIT Press).

Moore, G.E. (1922 [1907]) 'William James' "Pragmatism"'. In G.E. Moore (ed.) 1992, *Philosophical Studies*. London: Routledge and Kegan Paul, pp. 97–146.

Peirce C.S. (1935 [1877]) 'The Fixation of Belief'. In C. Hartshorne & P. Weiss (eds.) 1935, *Collected Papers* 5. Cambridge Mass: Harvard University Press, pp. 358–387.

Peirce C.S. (1935 [1878]) 'How to make our Ideas Clear'. In C. Hartshorne & P.Weiss (eds.) 1935, *Collected Papers* 5. Cambridge Mass: Harvard University Press, pp. 388–410.

Putnam, H. (2012) *Philosophy in an Age of Science* (Cambridge Mass: Harvard University Press).

Quine, W.V. (1953 [1951]) *From a Logical Point of View* (Cambridge Mass: Harvard University Press).

Quine, W.V. (1960) *Word and Object* (Cambridge Mass: MIT Press).

Quine, W.V. (1966 [1954]) *The Ways of Paradox* (New York: Random House).

Quine, W.V. (1969 [1968]) *Ontological Relativity and Other Essays* (New York: Columbia University Press).

Quine, W.V. (1975) 'On Empirically Equivalent Systems of the World', *Erkenntnis*, 9:313–328.

Quine, W.V. (1981) *Theories and Things* (Cambridge Mass: Harvard University Press).

Quine, W.V. (1981a) 'The Pragmatists' Place in Empiricism'. In R.J. Mulvaney & P.M. Zeltner (eds.) 1981, *Pragmatism: Its Sources and Prospects*. Columbia: University of South Carolina Press, pp. 21–39.

Quine, W.V. (1981b)[1978] 'Goodman's Ways of Worldmaking' *Theories and Things* (Cambridge MA: Harvard University Press):96–99

Quine, W.V. (1986) 'Autobiography'. In L.E. Hahn & P.A. Schilpp (eds.) 1986, *The Philosophy of W.V. Quine*. La Salle, Il: Open Court, pp. 3–46.

Quine, W.V. (1991) 'Two Dogmas in Retrospect', *Canadian Journal of Philosophy*, 21:265–274.

Russel, B. (1966 [1910]) *Philosophical Essays*, revised edition (London: GeorgeAllen and Unwin).

8

On Quine's Debt to Pragmatism: C.I. Lewis and the Pragmatic A Priori

Robert Sinclair

At the very end of his influential 'Two Dogmas of Empiricism', Quine famously emphasized that his rejection of the analytic-synthetic distinction resulted in a more 'thorough' pragmatism than that seen in the work of Rudolf Carnap and C.I. Lewis.[1] This remark has led many to assimilate Quine's work to the American pragmatist tradition, where he is often depicted as either continuing or reviving some of the main issues representative of that tradition.[2] Quine, however, remained ambivalent about this affiliation, explaining that he was only referencing Carnap's view concerning the pragmatic criteria involved in the choice of a linguistic framework for science, and recommending their extension to the whole of science (1991: 272). What he somewhat surprisingly forgets to mention is the influence of his teacher C.I. Lewis, who defended a form of 'Conceptual Pragmatism' in the 1930s when Quine was a graduate student at Harvard.[3] While the links between Quine and the 'classical' pragmatism of Peirce, James and Dewey are, I think, tenuous at best, I have earlier argued that it is precisely this connection to Lewis that serves as the main source of Quine's pragmatism (Sinclair, 2012). Here, I aim to further defend and elaborate on this claim by showing how Lewis's influence can be seen in several early episodes in Quine's philosophical development.[4]

Quine's epistemological views share many affinities with Lewis's conceptual pragmatism, where knowledge is conceived as a conceptual framework pragmatically revised in light of what future experience reveals. However, in this paper I will place special emphasis on the more specific core debt to pragmatism found in Quine's understanding and modification of Lewis's own central contribution to pragmatism, namely, his pragmatic conception of the a priori. We will see that Quine's early discussions of analyticity and the a priori endorse (if somewhat tacitly) Lewis's view of the a priori as a conceptual structure of our own making, where this structure is further seen as both analytic and extending into empirical science. However, in the process of defending the idea that the a priori be deemed analytic, Quine further suggests that the distinction between the a priori and the empirical is one of degree rather than kind. His use of the pragmatic conception of the a priori results in an increasing willingness to minimize the strictness of the analytic-synthetic distinction that foreshadows his later argument for the extended use of pragmatic criteria beyond the a priori into the empirical. Quine's gradual assimilation of Lewis's pragmatic a priori then results in his insistence that pragmatic criteria be extended to the justification of empirical claims more generally, something that Lewis himself would have rejected.

In order to support these conclusions, my discussion will focus on Quine's graduate work from the 1930s, which I will argue serves as a bridge from Lewis's conceptual pragmatism to Quine's reflections on Carnap's logical syntax project in his 1934 Carnap Lectures. These papers reveal a thorough understanding of Lewis's epistemology, notably its structural distinction between the conceptual and empirical components of human knowledge, the emphasis on the use of conceptual frameworks to interpret experience, and the further importance of the creative, pragmatic decision making involved in the choice of such a framework. They also indicate signs of Quine's reluctance to endorse a strict separation between the conceptual and the empirical elements of knowledge that anticipates his later rejection of the analytic-synthetic distinction. I then examine the first part of Quine's 1934 Carnap lectures in order to show how he marshals key elements in Lewis's pragmatist view of the a priori in his defense of a Carnap-like conception of philosophy as logical syntax. Quine's early graduate work helps to situate his approach in these lectures, especially the specific method he uses to capture the analytic and a priori status of accepted sentences. I then briefly consider relevant aspects of his later criticisms explaining how his use of the pragmatic a priori in his early work results in his questioning the epistemological importance of Lewis's emphasis on the distinct conceptual and empirical elements of human knowledge. What this reveals is the significance of Quine's use of Lewis's pragmatic a priori for the development of his later epistemological dismantling of the analytic-synthetic distinction. Quine's remark about his more 'thorough' pragmatism is then an apt description of these early episodes in his philosophical development, since his interpretation of the pragmatic a priori results in its extension into the empirical, opening up the possibility of synthetic claims becoming analytic. This liberalization of Lewis's treatment of the a priori will eventually lead to pragmatic criteria entering into the justification of empirical claims and Quine's further conclusion that a strict analytic-synthetic distinction is idle for epistemological purposes. In order to see this, we must first begin with an examination of Lewis's epistemology from his 1929 *Mind and the World Order*, especially his pragmatic conception of the a priori.

1   Lewis's pragmatic conception of the a priori

Lewis's general account of empirical knowledge highlights the way justified knowledge claims require both what is sensibly presented within experience (what Lewis calls the empirical 'given'), and the mind's own constructive activity (Lewis, 1929: 37; Dayton, 1995: 258). He then further distinguishes three main elements within empirical knowledge: the empirical given, the act of interpreting this given as an experience of something or other, and the concept through which we interpret the given by relating it to other empirical possibilities (Lewis, 1929: 230; Hunter, 2008). It is with the third conceptual component that Lewis introduces his novel account of the a priori, which he describes in these terms:

> the a priori is independent of experience not because it prescribes a form which experience must fit ... but precisely because it prescribes *nothing* to the content of experience. That only can be a priori which is true *no matter what*. What is anticipated is not the given, but our attitude toward it; it formulates an uncompelled initiative of mind, our categorial ways of acting. (1929: 197)

Lewis here conceives of the 'a priori' as comprised of those basic logical categories introduced in order to make sense of our sensory experience, and which further reflect those fundamental habits of thought that we have adopted in light of past attempts to render experience meaningful (1970a [1923]: 238). Importantly, our a priori system of concepts does not place any constraints on experience but simply shows our freely chosen commitment to classify and organize experience in ways that remain revisable in light of what future experience may reveal. Our system of knowledge is achieved through, as Lewis describes 'a process of trial and error' where he continues 'we have attempted to impose upon experience one interpretation or conceptual pattern after another, and guided by our practical success or failure, have settled down to that mode of construing it which accords best with our purposes and interests of action' (1970b [1926]: 251). The distinctive pragmatist nature of Lewis's account is then to be found in his emphasis on human knowledge as a creative activity where the conceptual elements used to make sense of experience are further rooted in various sorts of human interests and needs (Lewis, 1970 [1926]: 241; Hookway, 2008: 282).

With this pragmatist view of the a priori in place Lewis continues by highlighting its connection to analytic truth and analyticity. He explains that a priori truth emerges from the concepts themselves in two distinct ways (1929: 230–231). The first is most clearly seen in mathematics where this a priori truth involves elaboration of concepts in abstraction from any consideration of how they may apply to experience. With the second we witness the important role concepts play in empirical knowledge, since their application to experience reveals their status as 'predetermined principles of interpretation' that further serve as our criteria of reality in classifying and organizing experience. In both these ways, Lewis explains that truth is fixed independently of experience and simply represents the elaboration of the concept itself. Not surprisingly, it is here that he further identifies a priori truth with analyticity: '*The a priori is not a material truth, delimiting or delineating the content of experience as such, but is definitive or analytic in its nature*' (Lewis, 1929: 231, italics in the original). Lewis is quick to point out that while a priori principles are created by us and are susceptible to change, the choices involved in this process of creation are not arbitrary since, as we have seen, they must answer pragmatic criteria (1929: 237–238). Human interests and needs are either met or resisted in experience through the chosen conceptual system used to give order to experience. If those conceptual principles and criteria of interpretation fail to help us order and simplify our experience then they will be rejected in favor of another set of principles. The precise nature of this kind of conceptual change is important for understanding the Quine-Lewis connection and will be returned to below.

Lewis further argues that this a priori element in human knowledge carries over in a profound way to natural science. All order and scientific law depends on a prior ordering of experience, where we have seen that such a priori principles are human creations. Without such starting points to make sense of experience, we would remain faced with an unorganized, incomprehensible experiential mix. Lewis expands on this point in the following way: 'In every science there are fundamental laws which are a priori because they formulate just such definitive concepts or categorical tests by which alone investigation is possible' (1929: 254). This is further illustrated with Einstein's definition of simultaneity, one that Quine too uses in his later appropriation of Lewis's conception of the a priori. The issue here concerns determining whether two events happened at the same time, for example, lightning striking a railroad track at two places, A and B. Lewis further describes Einstein's attempt to give a definition of simultaneity that allows us to determine whether or not the lighting strikes happened at the same time. He supplies the required definition by sketching how someone properly positioned with appropriate visual aids could observe both A and B at the same time. Upon witnessing two flashes we could then say they are simultaneous. This 'definition' of simultaneity provides us with a clear way to make an empirical decision concerning whether this concept can be correctly applied or not. That it depends on light requiring the same amount of time to travel from A and B to the observer is, Einstein explains, a stipulation that we have freely chosen in order to arrive at our definition of 'simultaneity'.

It is not difficult to see why Lewis takes this scientific example as illustrating his pragmatic view of the a priori since Einstein's discussion highlights all of the main features of this account. For Lewis this definition counts as a clear case of an 'a priori stipulation' that further enables us to formulate 'definitive criteria' (1929: 256). Einstein freely chooses to define simultaneity in this way, that is, as collision of light at a midpoint between two sources. This definition then provides clear empirical application conditions for this concept that further enable the kind of ordering and interpretation needed to make sense of sensory experience. Lewis describes such definitions as 'laws which prescribe a certain behavior to whatever is thus named. Such definitions are a priori; only so can we enter upon the investigation by which further laws are sought' (1929: 256). The crucial point here concerns the logical priority of this a priori classification for making sense of experience at all and the further claim that this a priori classification is pragmatic highlighting 'the responsiveness of truth to human bent or need, and the fact that in some sense it is made by the mind' (1929: 271).

With this view in place, we can see that concepts stand as criteria for the classification of sensory experience that further enable the making of empirical judgments, and which then fit together within a larger system of classification. Lewis explains that applying any one of these concepts to any specific experience is probable only, but the further application of the larger system of concepts involves the choice of an abstract system, and this, according to Lewis, can only be determined through considerations of utility, stability and convenience (1929: 298–299). If future experience does not accord with the consequences of that concept, we will retract its application from the particular experience in question. The continued failure of individual concepts to successfully apply to experience may lead to readjustments to the overall conceptual system.

It is especially important to note the exact nature of this type of change to our conceptual system. Consider Lewis's own example of the difference between the two propositions 'All swans are birds' and 'All swans are white' (1929: 302–303). The first is an established definition that explains the meaning of the word 'swan'. If we thought that something was a swan and then discovered it was not a bird, we would no longer apply this concept and look for an alternative. Moreover, since this proposition is definitional for our use of the term 'swan', we can further note its status as an analytic truth. The second proposition is an empirical claim because the concept 'swan' does not imply any specific color further revealing that it can be falsified by what future experience may tell us about the color of swans. Empirical generalizations are then as Lewis describes, probable only, while a priori definitional claims can only be viewed as useful or not. In this latter case the needed replacement of one concept for a more useful one would not, strictly speaking, amount to a falsification of this concept's definition and its corresponding criteria of application. Lewis then explains:

> Definitions and their immediate consequences, analytic propositions generally, are necessarily true, true under all possible circumstances. Definitive is legislative because it is in some sense arbitrary... If experience were other than it is, the definition and its corresponding classification may be inconvenient, fantastic, or useless, but it could not be false. (1970 [1923]: 233; 1929: 239–240)

In this way, a system of concepts remains true in terms of the relations between the definitional meanings of its concepts, even if this system were to prove unhelpful for interpreting experience (Lewis, 1929: 268–270; Murphey, 2005: 159).

Summing up, we have seen Lewis defend the following features of his conception of the a priori:

1. Its pragmatic character, where our a priori system of classification shows a responsiveness of truth to human interest and is itself a human creation.
2. The a priori is deemed analytic and definitive.

3. Pragmatic a priori principles are present in natural science as seen in the case of Einstein's definition of simultaneity.
4. Strictly speaking, this a priori conceptual system is never falsified but remains true to its own internal semantic structure. Experience may reveal that it is no longer useful, but not false.

It is this epistemological view of the a priori that Quine encountered as a graduate student in the 1930s and while he comes to share Lewis's account of the use of conceptual systems for organizing experience, and their pragmatic adjustment in light of future experience, his eventual rejection of a strict conceptual-empirical distinction will result in the pragmatic choice of a conceptual system being extended to the justification of statements within that conceptual system itself. We can see the beginning of this influence and further disagreement with Lewis by looking at his unpublished graduate work from the 1930s.

2   Quine's early pragmatism

In the fall of 1930 Quine arrived at Harvard with his Oberlin BA in mathematics, which included an honors reading in mathematical philosophy but little previous training in philosophy. While his early student writings at Oberlin reflected a deep interest in the extension of scientific ideals across all areas of human inquiry, with a noticeable trace of empiricism, his study of the theory of knowledge was slight at best. In his first year at Harvard, Quine completed courses on Plato, Leibniz, Kant, and the Theory of Knowledge, with the last two being taught by Lewis. Having had little previous training in epistemology, Quine utilized the account outlined in Lewis's *Mind and the World Order* as the basis for his early reflections on this topic. His graduate papers demonstrate a thorough understanding of Lewis's pragmatic a priori, which we will see him use to experiment with the idea that we decide what should count as analytic within our evolving system of knowledge. Quine further considers the possibility of making the synthetic statements of empirical science analytic, an approach that develops into the later more precise method of definition he uses in his 1934 Carnap lectures. All of this points to Lewis's conceptual pragmatism having an important impact on Quine's early philosophical development, especially with regard to his understanding of the epistemological status of the analytic-synthetic distinction.[5]

An examination of Quine's unpublished graduate papers from this period provides further support for these claims. Because these papers are not widely available I will discuss them at some length and include some extensive quotes. Quine wrote two papers for Lewis entitled 'On the Validity of Singular Empirical Judgments' and 'Futurism and the Conceptual Pragmatist', where he discusses familiar Lewisian themes, including the significance of a distinction between the empirical given, and the conceptual frameworks that organize sensory experience, and the further claim that the application of any such system involves pragmatic standards reflective of our human interest in simplicity, utility and convenience: 'The *a priori* is a pragmatically devised instrument whose function is to aid in the control of future *given*. Insofar as a judgment involves the *a priori* network of concepts, therefore, the subject matter of the judgment is a subject matter whose very genesis was pragmatic'. He further emphasizes that this a priori system of concepts is revisable given pragmatic adjustments to experience: 'Man frames his concepts in that manner which promises most effectively to forward his interests; simplicity, frequent applicability, and applicability in important practical matters, are the prime considerations which, implicitly or explicitly, mould the bulk of our concepts'. The identification of the a priori and analytic truth is also explained: 'The a priori is occupied with the meanings of words, the connotations of concepts rather than their denotations; it is purely definitive, whence arises its apodictic validity.' Lastly, he demonstrates an understanding of Lewis's conception of the a priori as consisting of conceptual truths that are revised in terms of their retraction in favour of an alternative set of categorical conceptual commitments where such definitions are not falsified through experience: 'concepts are *logically* prior to the given in experience, in the sense that a given experience may or may not fall under a given concept, but no experience can ever falsify the definitive intension of a concept.'[6]

All of the central features of Lewis's pragmatic a priori are here on display. Quine's early graduate work exhibits his thorough understanding of the epistemological framework found in Lewis' *Mind and the World Order,* and he proceeds to address issues internal to this view of human knowledge, specifically dealing with technical concerns concerning how this pragmatist viewpoint can explain the validity or justification of empirical judgments.[7]

This can be seen with his emphasis on the role of a priori definitions in empirical science and the way he further addresses the issue of the relationship between conceptual systems and experience. Here, we see his agreement with Lewis concerning the importance of a priori definition for science:

The <u>a priori</u> is operative in our treatment of experience by providing and relating the concepts under which experience is to be subsumed ... certain so-called laws of physics or of any other discipline have this a priori and purely definitive character ... definitions of terms are necessary if one is know what he is talking about; and these definitions must ultimately constitute a framework of concepts which are interconnected by mutual definition.

Quine is here emphasizing Lewis's point concerning the logical priority of a priori principles for the interpretation and classification of experience and its further extension to fundamental scientific laws. We need, as he puts it, a set of clearly defined terms if we are to understand what we are talking about, and here he highlights the importance of a priori definitions in fulfilling this role. Quine now makes a further suggestion concerning how empirical claims can be made into a priori judgments in way that mirrors Lewis's treatment of Einstein's definition of simultaneity:

Clearly an empirical generalization can be replaced by an a priori judgment of a form closely related to it. Consider an empirical generalization of the form "Every A̲ has p̲." So fundamental might be the role of this "natural law" in subsequent investigations, that we might incorporate it into a conceptual structure by what is essentially a process of redefinition ... Whether or not a given empirical generalization will be thus definitionally hypostatized, will depend upon those pragmatic considerations which underlie our moulding of the a priori in general.[8]

Given the importance of having a precise set of principles through which to proceed with scientific inquiry, we may reach a point where an empirical claim becomes so central for subsequent investigations that we decide to integrate into our system through a process of redefinition. Quine briefly explains that assigning such importance to 'Every A is p' for example, would then involve redefining A by stipulating that in addition every A has p, where this states that every A exhibits the property p. This newly defined concept now becomes our criteria for assessing whether we have an A or not. Given the importance that we have now assigned to this definition, the discovery of an A that does not have the property p, would result in our refraining from categorizing it as an A. Through something like this process of redefinition, we can recognize how certain natural laws will take on an a priori or analytic status given their central scientific importance. Quine's suggestion here, is, as we have seen, reminiscent of Lewis's presentation of

Einstein's definition of simultaneity as a key instance of an a priori, pragmatic stipulation required for further scientific study. Quine does not mention this example until later, but his idea of making an empirical generalization an a priori law through redefinition is clearly connected to Lewis's ideas concerning the role a priori principles play in scientific inquiry.

This point resurfaces once again when Quine considers the modification of conceptual systems through contact with experience. In addition, we can note an early underdeveloped expression of Quine's reluctance to accept any sharp divide between the a priori and empirical. In his 'Concepts and Working Hypotheses', a paper written for Alfred North Whitehead rather than Lewis, Quine discusses the role of what he calls 'working hypotheses' in helping to increase the amount of simplicity found in our conceptual systems. In doing so he begins with the Kantian distinction between analytic and synthetic statements:

> This brings us to the distinction between concepts proper and the relations between them … These interconceptual connections may, in Kantian language, be either analytic or synthetic. The former would of course comprise at least two classes of relations: (1) intrinsic natures of the concepts themselves, and (2) those representing the exclusion of one concept formanother according to the same considerations. As to the synthetic relations between concepts, it appears moreover that the ideal of simplicity and unity in the system might be phrased as the idea of getting rid of all such relations—of carrying the concepts back to a single fountain-head concept in terms of which all the formerly synthetic relations would become analytic. So long as a department of study is active, however, its ideal has not been reached; and it is in the realm of these provisionally synthetic relations that the <u>working hypothesis</u> lies.

Here Quine suggests that in the course of structuring of our conceptual system on the basis of pragmatic standards of simplicity and systematic unity, we proceed to eliminate synthetic judgments in favor of their analytic counterparts hinting at an idea clearly connected to both his own and Lewis's suggestion that we stipulate an a priori definition and make an empirical claim into a law. However, in many areas of scientific study where hypotheses are still quite tentative, he recommends we refrain from making such claims analytic, a point that we will see him repeat in his Carnap lectures. Among the synthetic judgments in such scientific studies, Quine locates what he calls the 'working hypotheses' that face experience. It is here, he further explains, that error is to be located:

> As to synthetic relations, however, it remains in every case to be seen whether the items of experience in question could all be accommodated to an ideal system of concepts wherein these particular synthetic relations would have *become analytic*; and it is therefore in the assertion of those synthetic relations that error, falsification of experience, can arise. Thus it is that a working hypothesis is subject to change as recalcitrant items of experience are encountered.

But in this process of accounting for error and conceptual change, he appears to incorporate at least one important class of analytic or conceptual claims within our evolving empirical system of knowledge, whose 'validity', he claims, depends on the synthetic statements they are inferred from:

> If a recalcitrant item of experience, belonging to the field in question, should subsequently arise, modification somewhere in the system must take place; for it has been noted that a satisfactory conceptual system must accommodate every experience falling within the field. Thus it is that only the working hypothesis can stand which has endured without the emergence of any anomaly in the whole mass of experience since its inauguration. Analytic propositions are deduced on its basis … any violation of one of these by a subsequent experience would be a violation of the parent hypothesis. Failing any such violations, the system continues to grow; for other hypotheses have corresponding adventures, the successful ones remain and continue to beget analytic offspring, and groups of such hypotheses and their offspring unite in forming the basis for yet further analytic propositions.[9]

With the claim that analytic statements are inferred or deduced from empirical working hypotheses Quine suggests that they can be rejected on the basis of a more direct type of confrontation with experience. He further notes that there is a certain amount of latitude concerning where we wish to locate any possible 'error' in the conceptual system, and we may find it in the working hypothesis, or in our prior chosen set of concepts. This is similar to the type of holistic argument Quine would later use against a strict conceptual-empirical divide, and further suggests a type of modification to our conceptual a priori framework that

Lewis's sharper distinction between the conceptual and the empirical would not allow. With that distinction in place, we saw Lewis emphasize that empirical claims are probable only, while a priori statements remain definitional, being useful or not, but not strictly speaking false. By contrast Quine is suggesting that such the divide between analytic and synthetic statements is better thought of as one of degree. He further characterizes a type of analytic statement that *has* implications for experience through its inferential connections to empirical generalizations and suggests that it too can be viewed as probable and rejected on the same basis as synthetic claims. This coupled with his earlier suggestion concerning the a priori redefinition of central empirical claims illustrates Quine's willingness to minimize the strictness of the difference between the empirical and conceptual elements of human knowledge.

We are not, however, presented with the complete blurring of this distinction as found in Quine's later work, since he further recognizes a general type of conceptual or analytic statement involving the 'subsumption of one concept under another' and which he characterizes as distinct from empirical synthetic claims. Still, this critical perspective seen here will gradually develop into his more thorough rejection of any epistemological significance being given to this distinction. While Lewis uses the human made, pragmatic character of the a priori as a way to maintain a clear difference between the analytic and synthetic, Quine's initial resistance to the distinction develops from his use of this pragmatist conception of the a priori. We can further recognize the importance of this influence by taking a closer look at how these issues are further explored in Quine's discussion of analyticity in the first part of his 1934 Carnap lectures.

3 Quine's Carnap lectures and the pragmatic a priori

We have seen that Quine's graduate work provides a clear indication of the influence of Lewis's pragmatic a priori with Quine's exploratory use of the idea that we decide what claims are to be deemed analytic. This idea was then further used within the context of natural science, something Lewis also defends. These claims also play a prominent role in Quine's 'Lectures on Carnap' given at Harvard in the fall of 1934, which provide an enthusiastic defense of Carnap's Logical Syntax program.[10] My focus here will be the first lecture titled 'The A Priori'. Quine's aim in this lecture is to demonstrate how the a priori sentences of our language can be rendered analytically true by definition. He begins by discussing the nature of definition, and in the process outlines a method for demonstrating how a significant part of logic can be made true by definition. He concludes by addressing the question of how far we might extend this process to the remaining empirical part of our vocabulary. Overall he provides a defense of the view that the a priori should be construed as analytic, where he is specifically interested in showing how, through a process of definition, we can proceed to clarify the epistemological credentials of the a priori. In carrying out this aim, he devises a strategy to defend the analytic nature of the a priori that is based on the pragmatist idea that a priori claims are in an important sense, made by us. He further extends this thought to the case of empirical claims, which, on the basis of pragmatic criteria, can be made analytic. Here he looks to Einstein's definition of simultaneity as a case in point, the very same example used by Lewis to defend his conception of the a priori. Lastly, in discussing where we should stop this process of definition, he stresses the importance of pragmatic standards, claiming that revisions to our system conform to his own maxim of minimum mutilation.

Quine begins his first lecture with a preliminary characterization of analyticity, where analytic judgments are depicted as consequences of definitions. These are, he further explains, conventions governing the use of words with analytic claims being 'consequences of linguistic fiat' (Creath, 1990:2). After rejecting Kant's view of synthetic a priori judgments because of recent advances in the logical foundations of mathematics, he quotes, with approval, Lewis's identification of the analytic and the a priori. In order to further defend the claim that the a priori is coextensive with the analytic, he distinguishes between implicit and explicit definition.[11] Explicit definition simply introduces conventions for the abbreviation of terms. Quine's example is the word 'momentum', which is a linguistic convention introduced as an abbreviation for the expression 'mass times velocity' (Creath, 1990: 48). However, the implicit definition of a term specifies that a group of sentences containing that term are to be conventionally accepted as true. Unlike explicit definition it does not require any already defined terms and we can then render sentences true without relying on any other sentences. Quine notes how the use of definition plays little role in language use until our studies require more clarity and rigour. Through his use of implicit definition, he proceeds to offer a method for this type of more reflective process of definition, which following Creath, we can call the 'method of accepted sentences' (1987: 480). We will see that it is derived from Lewis's ideas.

Quine begins with our current accepted body of truths, somewhat surprisingly placing aside any concern with the distinction between the a priori and the empirical. He asks us to take some term K and consider all the accepted sentences that contain K. Quine's aim is to then define K so that all the accepted K sentences will be true under that definition. There is an infinite number of such accepted sentences, and so we must work from finite resources in establishing the required definitions. Quine further explains that some K-sentences will contain other words, say H, and there will be then a set of H sentences. We need to then decide whether to define such sentences under K or as sentences with the word H. Here, a technical distinction is introduced between vacuous and material appearances of a word: 'Any sentences which contains a word H...and remains unaffected in point of truth or falsity by all possible substitutions upon the word H...will be said to involve H vacuously' (Creath, 1990: 51). By contrast, the appearance of the term is material when its substitution does change the truth value of the sentence. Quine's example 'Within any class of two apples there is at least one apple' uses the word 'two' materially, but the word 'apple' vacuously. Noting this difference helps, since if an accepted sentence contains a term vacuously it will be more convenient to account for that sentence through defining words that occur materially. When more than one word occurs materially, the question of which word we should define in order to account for that sentence will be a more or less arbitrary choice, but one that, Quine emphasizes, is guided by pragmatic criteria such as convenience:

> Relatively to every concept, either individually or at wholesale, the priority of every concept must be favorably or unfavorably decided upon. In each case the choice of priority is conventional and arbitrary, and presumably to be guided by considerations of simplicity in the result. (Creath, 1990: 52)

On the basis of this suggestion, Quine's method of accepted sentences proceeds in the following way: priority is given first to logical concepts over both empirical and mathematical concepts, then within logical concepts themselves, it is 'neither-nor' that is given priority over all other logical notions. Given this strategy, defining 'neither-nor' will then involve accounting for 'neither-nor' sentences with no other word occurring materially. Quine continues by giving an example of such a sentence, and then by demonstrating with the help of the Sheffer stroke how we can formulate a finite set of rules that account for an infinite set of 'neither-nor' sentences (Creath 1990: 53–55).[12]

By adopting these rules we have in place an implicit definition of 'neither-nor', and in the process Quine claims that we have then shown: 'All accepted sentences materially involving only "neither-nor" become *analytic*: they become consequences merely of the linguistic conventions...governing the use of "neither-nor"' (Creath, 1990: 55). He further explains that other logical notions can be explicitly defined by using 'neither-nor' and others defined implicitly with the introduction of other conventions. The process can then extended to mathematical terms as seen in Whitehead's and Russell's *Principia Mathematica*, where, Quine claims, logical notions can be used to implicitly define all of pure mathematics.

It is here that he begins to wonder where we should stop this process. We can, he suggests, extend this method to include our use of empirical words such as 'event' or 'energy' and thereby make such sentences analytic: 'We could go on indefinitely in the same way, introducing one word after another, and providing in each definition for the derivation of all accepted sentences which materially involve the word there defined and preceding words but no others' (Creath, 1990: 61–62). Accounting for every word in the English language through such a procedure would yield the, perhaps surprising result, that every accepted English sentence can be made analytic, that is, they could all be derived from the conventions that we have established concerning the use of words.

This suggestion is clearly related to a point we saw in Quine's graduate work, where he stresses that an empirical generalization, can, on pragmatic grounds, be made analytic through a process of redefinition precisely because of its significance for subsequent scientific study. However, the possibility remains that we wish to still retain a line between the conventional and the empirical and here Quine proceeds to give some principled reasons for why we would stop this process of conventionalizing. The most important involves the revision of scientific hypotheses. New scientific discoveries require the revision of old empirical laws, and in carrying out such revisions we have some degree of control over where to make changes. Here we find an early expression of an idea often mentioned by Quine, where these choices are based on our tendency to disturb as little of our previous view as possible in order to be compatible with the additional demands of unity and simplicity of the system. This idea was in evidence in his graduate work as well, where in the attempt to 'accommodate' experience we mold our conceptual structure according to the same demands of simplicity and unity of system. Here, Quine is more explicit in extending this idea to revisions to the system, where these must adhere to pragmatic criteria, while preserving as much of our previous conceptual structure. Given this, it is useful to assign a 'provisional' non-analytic status to those principles we will be most interested in modifying if the need for revision occurs. If we made all empirical generalizations analytic, we would find ourselves constantly defining and redefining in an unnecessarily complicated way, but more importantly would lack any criterion for revising one definition as opposed to another. However, the demands of clarify require more precise definitions where it is we who decide which accepted sentences be made analytic: 'we will do best to render only such sentences analytic as we shall be most reluctant to revise when the demand arises for revision in one quarter or another' (Creath, 1990: 62). Not surprisingly then, Quine claims that we should proceed to keep logic and mathematics analytic as well as those parts of empirical science that we remain reluctant to revise.

It is here that he appeals to Einstein's definition of simultaneity as an example of an empirical concept that has been made analytic, the central case that Lewis also used in defending his pragmatic version of the a priori. He explains that Einstein defined the simultaneity of light emissions as meaning the collision of the light at a midpoint between the two sources (Creath, 1990: 64). Einstein thereby adopted a convention in order to resolve the question of the simultaneity of events. Both Quine and Lewis interpret this case as one where a decision has been made with regard to what concepts form part of the a priori or analytic defining principles of our system of scientific knowledge. Quine's use of this example is, I suggest, no accident but can be reasonably seen as derived from Lewis's own discussion.

Quine thus arrives at his main point that analytic claims are true by linguistic convention and it is we who decide which claims are to be made analytic: 'How we choose to frame our definitions is a matter of choice. Of our pre-definitionally accepted propositions, we may make certain ones analytic, or other ones instead, depending upon the course of definition adopted' (Creath, 1990: 64). This viewpoint is then extended to the a priori where Quine concludes that we should characterize the a priori as consisting of these analytic sentences:

> there are more and less firmly accepted sentences prior to any sophisticated system of thoroughgoing definition. The more firmly accepted sentences we choose to modify last, if at all, in the course of evolving and revamping our sciences in the face of new discoveries...These, if any, are the sentences to which the epithet "a priori" would have to apply. And we have seen...that it is *convenient* so to frame our

definitions as to make all these sentences analytic... But all this is a question only of how we choose to systematize our language... the doctrine that the *a priori* is analytic remains only a syntactic decision. (Creath, 1990: 65)

Quine's basic defense of the view that analytic statements be deemed a priori can, I suggest, be seen as utilizing central ideas from Lewis's pragmatic rendering of the a priori. The creative, decision making process that results in a choice concerning what we take as an analytic truth, and which is further deemed 'a priori' given our reluctance to revise such definitions, while clearly connected to Lewis's view, has been wedded to Quine's further reading of Carnap's syntactic depiction of philosophical claims. While it is then quite evident that this discussion of analyticity is heavily influenced by Lewis's pragmatist view of a priori classification it is doubtful that Lewis would accept Quine's further reduction of his view to a question of syntactic decision (Lewis, 1970c). Nevertheless, it is this epistemological understanding of the pragmatic function of the a priori that serves as the context through which Quine situates his own interpretation and defense of Carnap's logical syntax project.

4   Glancing ahead: The pragmatic a priori and 'Two Dogmas of Empiricism'

We have then clear evidence of the influence of Lewis's pragmatism in Quine's early discussions of analyticity and the a priori found in both his graduate work and 1934 Carnap Lectures. A more thorough account of how Quine's assimilation of Lewis's view results in his final critical stance on the analytic-synthetic distinction would need to include further discussions of Quine's 'Truth by Convention', Lewis's later account of analytic truth in *An Analysis of Knowledge and Valuation* (1946), as well as the triangular correspondence between Quine, Morton White and Nelson Goodman on the topic of analyticity in the late 1940s.[13] In lieu of such a discussion, this concluding section briefly examines how the ideas we have seen influence aspects of Quine's critical reflections in his 'Two Dogmas of Empiricism'.

Quine's famous critique of the distinction between the conceptual and empirical, results in his further rejection of a firm difference between the pragmatic criteria involved in the choice of a conceptual system, and the additional question of the evidential support of statements within a chosen conceptual framework. It was precisely with this denial that Quine offers his pragmatism as more thorough than either Carnap's or Lewis's similar positions. The historical connections defended here between Lewis and Quine over the pragmatic a priori suggest that when taken as a description of his own development this is a fitting assessment. In order to see why this is the case we need to revisit Quine's claim that his blurring of the conceptual-empirical divide results in a more thorough pragmatism.

The argument that is most relevant here is Quine's familiar criticism of the so-called second dogma of empiricism involving 'reductionism', the view which claims that all meaningful statements can be translated into statements about immediate experience (Quine, 1981a [1951]: 38). Quine further connects this view to the analytic-synthetic distinction in the following way: 'as long as it is taken to be significant in general to speak of the confirmation and infirmation of a statement, it seems significant to speak also of a limiting kind of statement which is vacuously confirmed, *ipso facto*, come what may; and such a statement is analytic' (1981a [1951]: 41). This reductionist view appears to provide a way to maintain a sharp distinction between conceptual and empirical statements, since empirical claims are precisely those that have their own separate set of supporting experiences, while analytic claims are those that are true as Quine says 'come what may', that is, they have no empirical consequences (1981a [1951]: 41).

Responding to this view Quine famously presents his holistic view of human knowledge, which we have also seen at work in both his graduate work and Carnap lectures. He explains that 'The totality of our so-called knowledge or beliefs...is a man-made fabric which impinges on experience only along the edges' and that as a result 'No particular experiences are linked with any particular statements in the interior of the field, except indirectly through considerations of equilibrium affecting the field as a whole' (1981a [1951]: 42–43). This view of knowledge as one large overarching system of statements exhibits Quine's further commitment to the claim that statements have implications for experience only when forming part of a larger number of statements. This more inclusive set of sentences, rather than any isolated on its own, is, as he says, the proper 'unit of empirical significance', and it is this latter set that then has implications for experience (1981a [1951]: 42). The empirical consequences of any conceptual structure or theory cannot be distinguished sentence by sentence but instead extend across the entire system of statements, or more modestly, across a significant part of this system. Once human knowledge is depicted as this kind of interlocking system, there is no general principled way to distinguish between the synthetic sentences with their own separate amount of empirical content, and the analytic sentences consisting of our a priori commitments to a system, which have no such content.[14] But this then challenges any position, like Lewis's, which marks a clear difference between those pragmatic considerations that inform our choice of an a priori conceptual system, and the further question of the empirical justification of statements within this system. Questions of simplicity, convenience and fruitfulness do not simply impact the choice of a conceptual system, as Lewis suggests, but are now claimed to play a role within the system itself contributing to the justification of its statements.

Here, Quine emphasizes that what justifies a statement in general is that it is part of theory that yields better predictions of sensory experience than any other theory. What it means to say 'better' here is, in part, to recognize that the theory is simpler, more fruitful, and easier to use. He then agrees with Lewis that pragmatic concerns impact our choice of a conceptual scheme or framework, but that such concerns apply *in addition* to the choice and justification of statements within that framework. The result is a more thorough application of those pragmatic interests than Lewis suggests because while they influence the choice of a framework prior to its empirical confirmation, they are further applied to the statements within the system itself as an integral part of our overall attempt to justify that framework or system.[15] In trying to provide the most accurate theory for predicting and understanding what happens in the world, pragmatic concerns will unavoidably enter into our attempt to square our overall theory to the available evidence and thus play a basic role in its justification and acceptance.[16]

Whatever the merits of this argument, my key concern here is to indicate the role Lewis's influence plays in its construction. The issue here turns on the status of the conceptual-empirical distinction within Lewis's epistemology, and the central role that it plays within that account. Lewis's general account of knowledge is designed to explain how the conceptual and empirical elements of knowledge can be brought together to form well supported, justified empirical statements. This account is further based on Lewis's understanding of the distinction between the a priori and the empirical where he gives that distinction the following epistemological significance: it allows us to distinguish those sentences that make a claim on reality and can then answer to questions of evidence and justification (empirical, synthetic claims) and those that do not make any such claim on reality (a priori, analytic claims) and can be maintained whatever experience may reveal (Hylton, 2002: 12). We earlier saw that it is precisely because such a priori definitive statements make no claim on reality they can be maintained as true from within their own structural relations, even when found to not usefully apply to experience. The a priori is, as Lewis remarks, 'true no matter what', and he further stresses the point in these terms: 'The dividing line between the *a priori* and the *a posteriori* is that between principles and definitive concepts which *can* be maintained in the face of all experience and those genuinely empirical generalizations which *might* prove flatly false' (1970a [1923]: 176).

Now, it is precisely this epistemological understanding of the analytic-synthetic distinction that serves as the context for Quine's early philosophical development, eventually resulting in his mature criticism of the second dogma that we just witnessed. Quine's graduate training demonstrated his understanding (and perhaps tacit endorsement) of Lewis's pragmatic a priori, which presents the a priori as a human creation. But on the basis of similar, if underdeveloped thoughts on holism, Quine begins to question Lewis's further claim just noted that our a priori system of concepts is never falsified through experience but remains true to their own definitive 'meanings'. At this early stage, he suggests something only fully developed much later, that the connections between our conceptual frameworks and sensory evidence show that even those statements far removed from any direct confrontation with experience (and seemingly a priori) are implicated in our attempt to formulate justified claims about the world. Lewis's definitive concepts no matter how distant from a direct confrontation with experience will then have experiential consequences that make some kind of indirect claim on reality. It is on the basis of this idea that Quine would later famously claim that holistic considerations indicate that any statement can be held true come what may, and not simply Lewis's 'definitive concepts', if we are willing to revise other statements in this overarching structure that may conflict with what experience reveals.

Moreover, we saw that Quine's first Carnap lecture uses Lewis's pragmatic conception of the a priori in order to develop a method that helps to establish that analytic claims be viewed as a priori. In the process, he emphasizes the way empirical claims can be made analytic through a process of redefinition that is based on our pragmatic concerns involving simplicity, systematic unity and convenience. It is we who then decide where to draw the line between the analytic and synthetic and ultimately what should count as 'a priori' given our reluctance to revise certain definitional claims. While none of this early work is explicitly critical of analyticity, it shares, I suggest, an increasing willingness to view the analytic-synthetic distinction as one of degree rather than kind. Lewis wants to affirm both the human made character of the a priori and the importance of marking a strict analytic-synthetic divide for epistemological purposes. My suggestion is that Quine's use of the former view gradually leads him to question the latter claim. It is with this tendency to minimize the strictness of this distinction that then opens up a possibility only fully exploited by Quine much later, namely, the extension of pragmatic criteria beyond the a priori to the empirical. These criteria are then viewed as playing a role in the justification of empirical claims, where Quine concludes that every stage of the attempt to interpret experience through a human-made conceptual system appeals to pragmatic considerations (1981a [1951]: 46). Quine's early appropriation of Lewis's pragmatic rendering of the a priori then results in a gradual liberalization of the analytic-synthetic distinction to the point where its epistemological significance is lost. Lastly, it is through his use of Lewis's emphasis on the way human choice determines what counts as a priori truth, which further facilitates this mature criticism and his own more thorough conception of pragmatism.

Notes

1. Here is the infamous remark: 'Carnap, Lewis, and others take a pragmatic stand on the question of choosing between language forms, scientific frameworks; but their pragmatism leaves off at the imagined boundary between the analytic and the synthetic. In repudiating such a boundary I espouse a more thorough pragmatism, (1981a [1951]: 46).
2. For this suggestion see Creath (1990), Glock (2003) and White (2002). Further references and useful critical discussion of Quine's place in the pragmatist tradition can be found in Godfrey-Smith (2014) and Koskinen and Pihlström (2006).
3. See Quine (1990: 292) where he mentions Lewis's influence.
4. For historical accounts that discuss Lewis's possible influence on Quine, see Kuklick (1977), Misak (2013), Murphey (1968), (2005) and Isaac (2005, 2012). Others who have mentioned this influence or discussed it in more detail include: Baldwin (2007, 2013), Davidson (1994, 2004), Hookway (2008), Hunter (2008), Hylton (2007), and Koskinen and Pihlström (2006).
5. See Murphey (2012) and Isaac (2005, 2012) for further historical remarks concerning Quine's early education, including relevant discussion of Quine's unpublished graduate papers. Quine's own account is found in his (1985: 82–86).
6. The first, second and fourth passages are taken from Quine's unpublished graduate essay, 'Futurism and the Conceptual Pragmatist', May 6, 1931, W. V. Quine Papers (MS Am 2587). Houghton Library, Harvard University. The lengthy third passage is from his 'On The Validity of Singular Empirical Judgements', March 17, 1931, W. V. Quine Papers (MS Am 2587). Houghton Library, Harvard University.

7. This is the main aim of his rather technical essay 'On The Validity of Singular Empirical Judgements', March 17, 1931, W. V. Quine Papers (MS Am 2587). Houghton Library, Harvard University.
8. The last two quotes are from Quine's 'On The Validity of Singular Empirical Judgements', March 17, 1931, W. V. Quine Papers (MS Am 2587). Houghton Library, Harvard University.
9. The last three quotes are from Quine's unpublished graduate essay 'Concepts and Working Hypotheses', March 10, 1931, W. V. Quine Papers (MS Am 2587). Houghton Library, Harvard University.
10. My account of these lectures is indebted to Creath (1987), and Hardcastle (unpublished). Further useful discussion is also found in Hylton (2001).
11. Quine would later call implicit definition 'postulation' and explicit definition simply 'definition'. For discussion of these changes as they relate to Quine's development see Creath (1987) and Hardcastle (unpublished).
12. For the technical details see Quine's discussion in Creath (1990: 50–57) and the accounts of the lectures given in Creath (1987), and Hardcastle (unpublished).
13. Some of these further episodes are discussed in Frost-Arnold (2011) and Isaac (2011). The correspondence between Quine, White and Goodman is found in White (1999).
14. See Quine (1981b: 26–27). The exception concerns observation sentences that do have their own empirical implications. Quine's later acceptance of the theory-ladenness of observation sentences will result in some modifications to this view (Quine, 2000).
15. Quine thus explains: 'The organizing role that was supposedly the role of the analytic sentences is now seen as shared by sentences generally, and the empirical content that was supposedly peculiar to synthetic sentences is now seen as diffused through the system' (1981b: 28).
16. This paragraph is indebted to the discussion found in Hylton (2002).

Bibliography

Baldwin, T. (2007) 'C.I. Lewis: Pragmatism and Analysis'. In M. Beaney (ed.) 2007, *The Analytic Turn*. New York: Routledge, pp. 178–195.

Baldwin, T. (2013) 'C. I. Lewis and the Analyticity Debate'. In E.H. Reck (ed.) 2013, *The Historical Turn in Analytic Philosophy*. Basingstoke: Palgrave Macmillan, pp. 201–227.

Creath, R. (1987) 'The Initial Reception of Carnap's Doctrine of Analyticity', *Nous* 21(4): 477–499.

Creath, R. (1990) *Dear Carnap, Dear Van* (Berkeley: University of California Press).

Davidson, D. (1994) 'On Quine's Philosophy', *Theoria*, 60(3):184–192.

Dayton, E. (1995) 'C. I. Lewis and The Given', *Transactions of the Charles S. Peirce Society*, 31(2): 254–284.

Davidson, D. (2004) *Problems of Rationality* (Oxford: Clarendon Press).

Frost-Arnold, G. (2011) 'Quine's Evolution from "Carnap's Disciple" to the Author of "Two Dogmas"', *HOPOS*, 1(2): 291–316.

Glock, H.J. (2003) *Quine and Davidson on Language, Thought and Reality* (Cambridge: Cambridge University Press).

Godfrey-Smith, P. (2014) 'Quine and Pragmatism'. In G. Harman & E. Lepore (eds.) 2014, *A Companion to Quine*. Malden: Wiley-Blackwell, pp. 54–68.

Hardcastle, G. 'Quine's 1934 "Lectures on Carnap"', (unpublished).

Hookway, C. (2008) 'Pragmatism and the Given: C.I. Lewis, Quine and Peirce'. In C. Misak (ed.) 2008, *The Oxford Handbook of American Philosophy*. Oxford: Oxford University Press, pp. 269–289.

Hunter, B. (2008) 'Clarence Irving Lewis', The Stanford Encyclopedia of Philosophy, E. N. Zalta (ed.) http://plato.stanford.edu/archives/fall2008/entries/lewis-ci/.

Hylton, P. (2001) '"The Defensible Province of Philosophy": Quine's 1934 Lectures on Carnap'. In J. Floyd & S. Shieh (eds.) 2001, *Future Pasts*. Oxford: Oxford University Press, pp. 257–275.

Hylton, P. (2002) 'Analyticity and Holism in Quine's Thought', *The Harvard Review of Philosophy*, 10(1): 11–26.

Hylton, P. (2007) *Quine* (New York: Routledge).

Issac, J. (2005) 'W. V. Quine and the Origins of Analytic Philosophy in the United States', *Modern Intellectual History*, 2(2): 205–234.

Issac, J. (2011) 'Missing Links: W.V. Quine, the Making of "Two Dogmas", and the Analytic Roots of Post-Analytic Philosophy', *History of European Ideas*, 37(3): 267–279.

Issac, J. (2012) *Working Knowledge* (Cambridge, Mass: Harvard University Press).

Koskinen, H.J. & Pihlström, S. (2006) 'Quine and Pragmatism', *Transactions of the Charles S. Peirce Society*, 42(3): 309–346.

Kuklick, B. (1977) *The Rise of American Philosophy* (Cambridge, Mass: Harvard University Press).

Lewis, C.I. (1929) *Mind and the World Order* (New York: Dover Publications).

Lewis, C.I. (1946) *An Analysis of Knowledge and Valuation* (La Salle, Chicago: Open Court).Lewis, C.I. (1970a [1923]) 'A Pragmatic Conception of the A Priori'. In J.D. Goheen & J.L. Mothershead (eds.) 1970, *Collected Papers of Clarence Irving Lewis*. Stanford: Stanford University Press pp. 231–239.

Lewis, C.I. (1970b [1926]) 'The Pragmatic Element in Knowledge'. In J.D. Goheen & J.L. Mothershead (eds.) 1970, *Collected Papers of Clarence Irving Lewis*. Stanford: Stanford University Press, pp. 240–257.

Lewis, C.I. (1970c) 'Logical Positivism and Pragmatism'. In J.D. Goheen & J.L. Mothershead (eds.) 1970, *Collected Papers of Clarence Irving Lewis*. Stanford: Stanford University Press, pp. 92–112.

Misak, C. (2013) *The American Pragmatists* (Oxford: Oxford University Press).

Murphey, M. (1968) 'Kant's Children: The Cambridge Pragmatists', *Transactions of the Charles S. Peirce Society*, 4(1): 3–33.

Murphey, M. (2005) *C. I. Lewis, The Last Great Pragmatist* (Albany, N.Y.: SUNY Press).

Murphey, M. (2012) *The Development of Quine's Philosophy* (Dordrecht: Springer).

Quine, W.V. (1931a) 'Concepts and Working Hypotheses', W. V. Quine Papers (MS Am 2587). Houghton Library, Harvard University, (unpublished essay).

Quine, W.V. (1931b) 'On The Validity of Singular Empirical Judgements', W. V. Quine Papers (MS Am 2587). Houghton Library, Harvard University, (unpublished essay).

Quine, W.V. (1931c) 'Futurism and the Conceptual Pragmatist', W. V. Quine Papers (MS Am 2587). Houghton Library, Harvard University, (unpublished essay).

Quine, W.V. (1981a [1951]) 'Two Dogmas of Empiricism' in From a Logical Point of View (Cambridge: Harvard University Press) 20–46.

Quine, W.V. (1981b) 'The Pragmatist's Place in Empiricism'. In R.J. Mulvaney & P.M. Zeltner (eds.) 1981, *Pragmatism: Its Sources and Prospects.* Columbia: University of South Carolina Press, pp. 23–39.

Quine, W.V. (1985) *The Time of My Life* (Cambridge, Mass: MIT Press).

Quine, W.V. (1990) 'Comments on Parsons'. In R. Gibson & R. Barrett (eds.) 1990, *Perspectives on Quine.* Oxford: Blackwell, pp. 292 .

Quine, W.V. (1991) 'Two Dogmas in Retrospect', *Canadian Journal of Philosophy*, 21(3): 265–274.

Quine, W.V. (2000) 'I, You and It: An Epistemological Triangle'. In A. Orenstein & P. Kotatko (eds.) 2000, *Knowledge, Language and Logic: Questions for Quine.* Dordrecht: Kluwer, pp. 1–6. Reprinted in Quine, (2008).

Quine, W.V. (2008) *Confessions of a Confirmed Extensionalist and Other Essays* (Cambridge, Mass: Harvard University Press).

Sinclair, R. (2012) 'Quine and Conceptual Pragmatism', *Transactions of the C.S. Peirce Society*, 48(3): 335–355.

White, M. (1999) *A Philosopher's Story* (University Park: The Pennsylvania State University Press).

White, M. (2002) *A Philosophy of Culture: The Scope of Holistic Pragmatism* (Princeton, N.J.: Princeton University Press).

W. V. Quine Papers (MS Am 2587). Houghton Library, Harvard University.

Part IV
Understanding Quine

9
Quine's Philosophies of Language

Peter Hylton

Years ago, I was invited to write an essay on Quine's philosophy of language. I thought about this for a while. Quine has a great deal to say about language, but his remarks are of quite varying kinds. I could not see how they cohered into a single project that deserved the name 'Quine's Philosophy of Language'. So I did not write the essay.

Now I think that there was good reason not to write it: there is no one project which is Quine's philosophy of language. Hence the use of the plural in the second word of my title. My claim is that in Quine's work there are two quite distinct enterprises each of which has some claim to be called 'Quine's philosophy of language'. To put it another way: what other philosophers might call 'philosophy of language' divides, in Quine's thought, into two quite different enterprises. An example of someone who does *not* hold the kind of view that I am attributing to Quine – someone for whom the two enterprises are unified – is provided by Russell in the first two decades, or a little less, of the twentieth century. So I will approach the topic by first discussing the relevant views of Russell; then I will discuss Quine's two projects and how they differ.

1 Russell's logically perfect language

I begin with the idea of a logically perfect language, as Russell articulates it in the 1918 lectures published under the title 'The Philosophy of Logical Atomism' (Russell, 1986 [1918]). His logically perfect language (LPL, as I shall sometimes say) is a language in which a true sentence perfectly reflects the fact that makes it true (so the sentence, and the fact have the same structure). So the language is, as I shall say, *ontologically significant*. If a sentence of that language is true, it shows what entities there are which make it true, namely the entities named by the simple terms of the sentence. (Presumably also the entities which must be in the range of variables of the quantifiers in the sentence, if it contains quantifiers. Russell, however, says little about this, perhaps because he has no understanding of generality which he finds satisfactory.)[1]

It is worth emphasizing that, on Russell's account, ordinary language is *not* in this sense ontologically significant. Many true sentences of ordinary language contain singular terms which, Russell claims, do not refer to anything. More accurately: many sentences of ordinary language correspond to facts which do not contain an object or entity which corresponds to the singular term or terms which the sentences contain. As a guide to ontology, such sentences are thus misleading. Russell takes the sentence 'Piccadilly is a pleasant street' as an example. He clearly counts this sentence as true. If the language in which it is stated were ontologically significant in my sense, then its truth would imply that there is an entity named by the word 'Piccadilly', and that the sentence is about that entity. But that is not Russell's view. To reveal what that sentence is really about, what it really says, it must be analyzed, and the analysis shows that the ordinary language sentence is misleading:

> Suppose you made any statement about Piccadilly, such as "Piccadilly is a pleasant street." If you analyze a statement of that sort correctly, I believe that you will find that the fact corresponding to your statement does not contain any constituent corresponding to the word "Piccadilly". The word "Piccadilly" will form part of many significant propositions, but the facts corresponding to those propositions do not contain any single constituent, whether simple or complex, corresponding to the word "Piccadilly". That is to say, if you take language as a guide in your analysis of the fact expressed, you will be led astray in a statement of that sort. (Russell, 1986 [1918]: 170)

The analysis will reveal that the fact expressed by the sentence as uttered by Russell is made up of things quite different from streets, as we ordinarily conceive of them: most obviously, it is made up of sense-data, universals, and propositional functions.

What Russell says here about Piccadilly (or, more accurately, about the word 'Piccadilly') holds, in his view, for a very large number of (alleged) entities. In particular, it holds, as he says, for all 'apparently complex entities'. In other words, it holds for all the things that we ordinarily talk about or think about:

all the ordinary objects of daily life are apparently complex entities: such things as tables and chairs, loaves and fishes, persons and principalities and powers – they are all on the face of it complex entities. All of the kinds of things to which we habitually give proper names are on the face of them complex entities. (Russell, 1986 [1918]: 170)

These things are only *apparently* complex entities not because they are really simple but because – to put the point paradoxically – they are really not entities at all. As Russell says: 'For my part, I do not believe in complex entities of this kind' (Russell, 1986: 170). Such alleged complex entities are what Russell sometimes calls 'logical fictions' or 'logical constructions'.[2] What he means by this is best explained by speaking of the singular terms which appear (misleadingly) to refer to such objects, e.g. the word 'Piccadilly'. Sentences containing this word (and countless other singular terms) are analyzable, and in their fully analyzed forms they do not contain that word. Hence such sentences may be true (and some are true) even though there is no such entity as Piccadilly.

So ordinary language, in Russell's view, is not ontologically significant: a true sentence of ordinary language is not a reliable guide to reality. Russell's logically perfect language, by contrast, has exactly that feature. He puts it like this:

In a logically perfect language the words in a proposition [here meaning a sentence] would correspond one by one with the components of the corresponding fact.... In a logically perfect language, there will be one word and no more for every simple object, and everything that is not simple will be expressed by a combination of words, by a combination derived, of course, from the words for the simple things that enter in, one word for each simple component. (Russell, 1986 [1918]: 176)

So we can draw ontological conclusions directly from the true sentences of the LPL: a true sentence of that language accurately reflects the corresponding fact, unlike sentences of ordinary language.

Russell's logically perfect language also has another feature which has to do not with ontology, but rather with how we understand language. Every meaningful sentence (of any language) expresses a thought, or a judgment as Russell often says.[3] A given person's judgments are made up of entities (both particulars and universals) to which that person stands in a direct and immediate epistemological relation – entities with which he or she is acquainted, as Russell usually puts it. That, in Russell's account, is what makes it possible to understand a sentence, or to engage in propositional thought at all. Sentences of ordinary language usually disguise the real form and structure of the judgments that they express, and thus do not reveal with which entities we must acquainted if we are to understand the sentence. But for every meaningful sentence of ordinary language there is sentence which expresses it in a way that *does* accurately reflect the structure of the thought, and this is a sentence of the LPL. That sentence will conform to an epistemological requirement: each simple symbol of the sentence will correspond to an entity with which a person must be acquainted in order to understand the sentence.

Another way to make this point is in terms of analysis: we begin with an unanalyzed sentence of ordinary language and, through the process of analysis, find a fully analyzed sentence which reveals what is involved in understanding the ordinary language sentence. The ordinary language sentence and the fully analyzed sentence say the same thing; they are synonymous in a quite precise sense. The fully analyzed sentences make up the LPL. What constrains the process of analysis? In other words, when is it complete? It is of no use here to say that analysis is complete when we have obtained a fully-analyzed sentence, for the question is about what it is for a sentence to be fully analyzed, rather than susceptible of further analysis. Russell's answer is that a sentence is fully analyzed when all the terms in it correspond to entities with which we are acquainted. The criterion for when the process of analysis is complete, and hence also for a given sentence's being part of the LPL, is thus epistemological. (Since different people are acquainted with different entities we should, strictly, use the first-person singular rather than the first person plural: a sentence of *mine* is fully analyzed when all its terms stand for entities with which *I* am acquainted. Russell, however, always uses the first-person plural and I shall generally follow him in this.)

Implicit here is a model of what it is to understand a sentence. A necessary condition is that one understand each of the simple terms in the sentence: 'the components of a proposition are the symbols we must understand in order to understand the proposition' (Russell, 1986: 175). And to understand such a symbol one must be acquainted with the entity for which it stands. Russell takes the word 'red', as an example. He says: 'The word "red" can only be understood through acquaintance with the object' (Russell, 1986: 174). The point is a general one: to understand any term one must be acquainted with the relevant entity. Russell speaks of that entity as the 'meaning' of the symbol. Hence, as he says: 'All analysis...always depends...upon direct acquaintance with the objects which are the meanings of certain symbols' (Russell, 1986: 173).

This view is by no means new with the Lectures. It is explicit in *Problems of Philosophy*, written six or seven years before he gave the Lectures on the Philosophy of Logical Atomism. In his discussion of the analysis of definite descriptions, Russell makes it clear that the result of the analysis is a sentence which, unlike the original unanalyzed sentence, accurately and explicitly expresses the thought concerned: 'the thought in the mind of a person using a proper name correctly can generally only be expressed explicitly if we replace the proper name by a description' (Russell, 1999 [1912]: 37). He also makes it clear that we must be acquainted with all the constituents of the propositions which we can understand: '*Every proposition which we can understand must be composed wholly of constituents with which we are acquainted*' (Russell, 1999 [1912]: 58, emphasis in the original). Here Russell is using the word 'proposition' for abstract entities, rather than for sentences, as he (mostly) does in the Lectures. So it is the constituents of those abstract entities with which we must be acquainted. But the point is the same: to understand a sentence, we must be acquainted with the entities for which the simple terms of the sentence stand. So, again, the process of analysis is not complete until we obtain a sentence in which every term corresponds to an object with which we are acquainted.

So Russell's view is that I can only understand a sentence if I am acquainted with all the constituents of the thought or (in the earlier idiom, the proposition) which it expresses. Sentences in ordinary language, however, typically do not reveal what those constituents are. For that purpose, analysis is required: the fully analyzed sentence *does* reveal the constituents with which one must be acquainted if one is to understand the sentence. (That it do so is, indeed, the criterion already given for the analysis's being complete: analysis is only complete when we have obtained a sentence in which every term corresponds to an entity with which we are acquainted.) So Russell's account of understanding, i.e. of propositional thought, does not apply directly to sentences of ordinary language, many of which contain terms (such as 'Piccadilly') which, according to Russell's 1918 view, do not stand for any entity at all. The account of understanding in terms of acquaintance, however, does apply directly to fully analyzed sentences, in virtue of the epistemological constraint on analysis. So every meaningful sentence expresses a thought, but sentences of ordinary language almost always express the thought in a misleading or inaccurate way. For every thought, however, there is a fully analyzed sentence which expresses that thought in an accurate and perspicuous way, and shows how we are able to think it (it shows which entities we must be acquainted with in order to have that thought). These fully analyzed sentences make up the LPL.

The LPL is thus not a language wholly distinct from ordinary language; rather, it underlies ordinary language, and gives meaning to its sentences. A sentence of ordinary language is meaningful only if it expresses a thought which is accurately expressed as a sentence of the (given person's) LPL. So we can say that the LPL is primary from the point of view of how language is understood (by a given person), and how it comes to be meaningful (for that person): Our earlier discussion showed that the LPL is also primary from the point of view of ontology. A sentence of that language reveals the claims that it is really making about the world, what entities in what arrangements there must be in order for the sentence to be true.

Now the point I wish to emphasize is that for Russell these two aspects of language, what might be called the ontological aspect and the aspect of understanding, are unified. The structure of a sentence of the LPL (though not of a typical sentence of our ordinary language) *both* reflects the fact which makes it true, or would make it true, *and* reflects the thought which the sentence expresses, and thus also reflects the way in which we understand the sentence. I have focused on this unified treatment of the two aspects as it appears in Russell's 1918 discussion, but I have also noted that it is not new at that stage. In fact, though I shall not attempt to show this here, the unification of the two aspects characterizes the view of propositions which Moore first articulates in 1898 in opposition to Bradley, and which Russell adopts shortly thereafter. Russell's views of language and propositional thought shift significantly throughout the period 1900–1918 but, with the exception of a fairly brief period, they always allow for this kind of unified treatment of the two issues.[4]

2   Quine on ontology

So far I have distinguished two aspects of language, one aspect concerned with ontology and a second aspect concerned with how we understand language. In Russell, as I emphasized, these aspects are very closely unified: a sentence of the LPL, a fully analyzed sentence, shows *both* the entities with which one must be acquainted in order to understand the sentence *and* the entities which make up the fact which makes the sentence true, if it is true. Now the point I want to make about Quine can be encapsulated by saying that he is concerned with each of these two aspects of language, but that he treats them in quite different ways. Each aspect corresponds to a Quinean project which might fit under the heading 'philosophy of language', but the two aspects correspond to projects which in Quine's thought are separate and quite different.

I start with the ontological aspect of language. To begin with, we should note that ontology is a central concern of Quine's. The seventh and final chapter of *Word and Object* is entitled 'Ontic Decision'. In that chapter, Quine builds on earlier discussions to put forward substantive views as to what does and what does not exist; this is clearly of a piece with the idea that 'limning the true and ultimate structure of reality' is among the philosopher's tasks (Quine, 1960: 221). The concern with ontology receives less emphasis in Quine's later work, but is always present.

Language is central to Quine's view of how our ontological endeavours should proceed. For Quine, however, as for Russell, it is not ordinary language that plays this role. Quine's claim here is not that it is hard to discover the ontological implications of a given body of sentences of ordinary language; he claims, rather, that there is nothing to be discovered. In spite of some clear cases, the question is too vague for there to be fact of the matter:

> Bodies are assumed, yes; they are the things, first and foremost. Beyond them there is a succession of dwindling analogies...there is no purpose in trying to mark an ontological limit to the dwindling parallelism...a fenced ontology is just not implicit in ordinary language. The idea of a boundary between being and non-being is a philosophical idea, an idea of technical science in a broad sense. (Quine, 1981: 9)

Ontology is not answerable to ordinary language. To evaluate the ontology of a theory, we need to regiment it in canonical notation. The framework of this language is first-order logic with identity. There is no claim that this framework or canonical notation itself underlies or is in any sense already implicit in ordinary language; to the contrary, the use of the framework is justified by the clarity and simplicity which it brings to our theory as a whole. (An important aspect of this clarity is precisely that it makes possible a clear and definite ontology.) Quine builds on earlier work (especially by Russell) to show that this framework can encompass much more than might appear at first sight. An important example here is that he adapts Russell's theory of descriptions to show that we can achieve the effect of a language with function-signs, or singular terms of any kind, without assuming them as primitive parts of the language. One result of this, Quine claims, is that all that needs to be added to the framework of first-order logic are predicates; in particular, those predicates required for the formulation of the best available knowledge. When our knowledge is regimented in the resulting notation, the result is regimented theory: our best overall knowledge formulated as clearly and simply as possible.

Speaking of canonical notation, Quine claims that 'all traits of reality worthy of the name can be set down in an idiom of this austere form if in any idiom' (Quine, 1960: 228). The qualification 'worthy of the name' is worth noting. Some of the claims we make in ordinary life are, in Quine's view, not sufficiently clear and objective to be worthy of a place in regimented theory. The idioms characteristic of such claims are, accordingly, excluded from canonical notation, with no more or less equivalent idioms to take their place. One well-known example is the idioms of propositional attitude and, in particular of indirect quotation, where we do not give someone's actual words but rather give a sentence which we take to be equivalent to it, or close enough. Quine accepts that we need such an idiom in ordinary life; it is not, he says, 'humanly dispensable' (Quine, 1960: 218). For Quine, however, this does not justify including it in canonical notation:

> In general the underlying methodology of the idioms of propositional attitude contrasts strikingly with the spirit of objective science…. An indirect quotation we can usually expect to rate only as better or worse, more or less faithful, and we cannot hope even for a strict standard of more and less; what is involved is evaluation, relative to special purposes, of an essentially dramatic act. (Quine, 1960: 218f.)

Similar comments apply to other cases, for example subjunctive conditionals (Quine, 1960: 222f.). One of the goals of regimenting theory is to 'enhance objectivity' (Quine, 1976 [1957]: 235) and the excluded idioms give rise to claims which do not serve this goal.[5]

Idioms which are excluded from Quine's canonical notation because they do not enhance objectivity are not thereby said to be meaningless.[6] Canonical notation, that is to say, does not exhaust meaningful language; he says explicitly that scientific language is 'a splinter of ordinary language, not a substitute.'(Quine, 1976 [1957]: 236). Quine, as we saw, is even willing to say that one such idiom is not 'humanly dispensable'. Here there is a clear contrast between his canonical notation and Russell's LPL. For every meaningful sentence of ordinary language, or at least for every particular meaningful utterance of a sentence, there is a sentence of the LPL which has the same content and makes the same claim.[7]

To return to Quine on ontology: given an overall understanding of what regimented theory would look like, we can draw ontological conclusions. Since there are no singular terms in canonical notation, all that is relevant here is what objects must be in the range of the variables in order to make the sentences of the theory come out true.

Actually coming up with regimented theory, reformulating the totality our knowledge so that it fits into Quine's schema, may well not be feasible even in principle, and certainly the task would take more work than is ever in fact going to be devoted to it. So regimented theory, the complete finished object, is not something which will ever exist. It is to this extent an idealization, though not as distant from our ordinary uses of language as Russell's LPL. Nevertheless, Quine claims that we can say, in broad outline, what would and would not find a place in regimented theory. We might think of such claims as the result of a thought-experiment, an answer to the question: what would our knowledge look like if it were reformulated along the lines indicated?

It is worth emphasizing that in regimenting theory the aim is to maximize the simplicity and clarity of our knowledge *as a whole*. If we focus on a single issue then we may be led to results which simplify the treatment of that issue but lead to excessive complications in the overall theory. An example here might be the treatment of statements of propositional attitude – statements of the form 'A believes that *p*', 'A doubts whether *p*', 'A wonders whether *p*', and so on. If we consider how to accommodate such statements, and take that question in isolation from others, then we may be led to the view that the best course is to postulate propositions as entities towards which a subject may have any one of a number of attitudes: believing it, doubting it, and so on. But accepting propositions would lead to other sorts of complications and inconveniences, or so Quine claims. In particular, he claims that it would leave us with a wide range of questions for which we have no clear answers, because there are no clear criteria for when two sentences express the same proposition. So while postulating propositions might have local advantages, so to speak, for our treatment of statements of propositional attitudes, it would have global drawbacks: regimented theory as a whole would be less clear, simple, and efficacious with such postulation than without. For this reason Quine does not postulate propositions.

Quine claims, indeed, that the only objects that must be within the range of the variables of regimented theory are sets and physical objects.

He attempts to show how our knowledge, or the parts of our knowledge which need to be taken seriously as embodying objective information about the world, can be formulated without quantifying over other kinds of alleged entities. The example of propositions has just been mentioned. The same holds for properties (attributes, as Quine usually calls them), for sense data, and for facts. Likewise for states of mind except insofar as they can be identified with the state of a corresponding body.

Quine's conclusion from this is ontological: alleged entities not within the range of the variables of regimented theory do not exist. Clarification and simplification of regimented theory has ontological consequences:

> Each reduction that we make in the variety of constituent construc-
> tions needed in building the sentences of the language of science is a
> simplification in the structure of the inclusive conceptual scheme of
> science ... The quest of a simplest, clearest overall pattern of canonical
> notation is not to be distinguished from a quest of ultimate categories,
> a limning of the most general traits of reality. (Quine, 1960:161)

Quine's regimented theory, like Russell's LPL, is a language with onto-logical significance.

In his efforts to clarify and simplify our theory, Quine expends consid-erable ingenuity to show how various idioms can be avoided, and how the theories which employ them can be reformulated. Two examples of this kind of maneuver have already been mentioned. One is his use of the technique of Russell's theory of descriptions to show how a language lacking singular terms can have the same expressive power (or as much as is required) as a language with singular terms. By doing this, we avoid potentially troubling questions such as how it can be true that Pegasus, say, does not exist, when there is no object for this statement to be true of. A second example already mentioned concerns statements of propo-sitional attitude; Quine claims that we can understand such statements simply in terms of persons and sentences, and that the considerable *prima facie* problems with such an approach can be overcome. A third example, from many that might be added to this list, is Quine's adoption of a central insight of logicism so as to avoid assuming numbers as enti-ties in their own rights, making do instead with sets of certain kinds.

In these and other ways, Quine philosophizes about language with a motivation that is partly ontological, and corresponds strikingly with ways that Russell sometimes discusses similar issues. All of this, and much more, makes up the first of the two Quinean projects that I want to distinguish.

3 Quine on the understanding of language

The second aspect of language that I distinguished concerns how we understand language. One might at first think that this is not a Quinean project at all. By his standards, the word 'understanding', like the words 'knowledge' and 'meaning', is insufficiently clear for use in a subject that aspires, as he thinks philosophy should, to scientific standards of clarity, rigour, and evidence. The only place where he puts any weight on the word at all is the 1975 essay 'Mind and Verbal Dispositions'. There he does explicitly suggest a behavioral account of understanding for certain simple occasion sentences, those whose truth-value varies from time to time. (The example he uses is 'This is red'.) He doubts, however, that such an account can be extended to sentences in general:

> Perhaps the very notion of understanding, as applied to single standing sentences [i.e. those true or false once for all, rather than true on some occasions and false on others], simply cannot be explicated in terms of behavioral dispositions. Perhaps therefore it is simply an untenable notion, notwithstanding our intuitive predilections. (Quine, 1975a: 89).

Given that Quine is inclined to think that understanding is 'simply an untenable notion', we can hardly attribute to him, without qualification, the project of giving an account of the understanding of language.

There is, however, a Quinean project which comes close enough for our purposes to the project of accounting for our understanding of language. Quine is concerned to show that it is possible to give an austerely naturalistic account of how we come to have the knowledge that we have. Most of what he does to show this consists of sketching an account of how cognitive language is acquired, or at least of how it might be acquired. Quine would perhaps be reluctant to describe this enterprise as showing how an *understanding* of cognitive language is acquired, because the word 'understanding' has implications which he does not accept, and because he does not think that there is much to say about what it is to understand even a modestly theoretical sentence. Nevertheless: an account of how cognitive language is acquired is, as far as it goes, also an account of what there is to be acquired, and thus of what understanding the language consists in, insofar as the word 'understanding' picks out a genuine feature of the world. As implied by the passage just quoted, what we are offered may in some cases fall well short of what we are, pre-theoretically, inclined to think such an account should include. Nevertheless, it will, presumably, show us what there really is to the idea of understanding, show us the real phenomena that lead us to talk of understanding.

In some cases, what Quine offers us does come very close to satisfying the pre-theoretical demands that we make on the vague idea of understanding. This is most obvious in the case of what Quine calls observation sentences, such as his example 'This is red'. According to the picture which Quine usually presents, what the infant has to acquire, in order to understand this sentence, is a relatively straightforward disposition. This is the disposition to respond to the relevant question ('Is this red?') with assent if she (the infant) is receiving neural intake within the appropriate range at the relevant time, and to dissent if not. (For the sake of simplicity, I shall ignore neural intake which leads neither to assent nor to dissent; for the same reason, I shall mostly just speak of assent, leaving dissent as understood.)

Acquiring this relatively straightforward disposition does not, however, suffice for full adult mastery of the sentence. For this sentence, as for other sentences which might seem to be purely observational, current neural intake does not in fact determine assent. Perhaps I assent to 'This is red' when I unwittingly observe a red light-source illuminating what is, in the absence of light from that source, a white surface. Had I been shown the set-up ahead of time, I would not have assented – even with the same neural intake that does in fact lead me to do so. So, contrary to the picture that Quine usually sketches, assent does not depend solely on current neural intake, even for a sentence as simple and as close to observation as 'This is red'. The rest of one's beliefs – one's theory, as Quine says – may also play a role. Since Quine thinks of that theory as embodied in language, this means that we cannot account for the understanding of even a very simple sentence without invoking other sentences, which would in turn require us to invoke yet others, and so on.

So the acquisition of the relatively straightforward disposition is not all that is required for full adult mastery of the use of the sentence: the adult withholds assent under certain circumstances, in spite of receiving appropriate neural intake. But still, for a sentence such as 'This is red', the infant who acquires the relatively straightforward disposition comes *very close* to acquiring full mastery of the use of the sentence, because occasions when her use will deviate from that of the adult will be very rare. Almost always, when things look red, they are. So the sentence 'This is red' may not, strictly speaking, fully meet Quine's criteria for being an observation sentence, but it comes very close. Treating observationality as a matter of degree, as Quine sometimes suggests (see e.g. Quine, 1960: 44), we may say that sentences like 'This is red' are very highly observational. Quine's view does not require that any particular sentence have this status, but it does require that some sentences do, for they provide the child's way into language.

Acquiring a disposition to assent to a highly observational sentence under appropriate neural intake is not all there is to being able to use the sentence as an adult can. But acquiring that disposition is a first step, on the basis of which other parts of the language can be learned. Eventually, taking advantage of this later learning, the child will, if all goes well, acquire a far more complicated disposition. Once she has acquired that more complicated disposition she will, on rare occasions, join the adults in refraining from assent even while receiving appropriate neural intake. Phrasing the point in terms of understanding: the child who acquires the relatively straightforward disposition does not thereby acquire a full understanding of the sentence. But she does acquire a partial understanding which comes close to a full understanding, and she may use that partial understanding to acquire an understanding (again, perhaps only partial to begin with) of other parts of language, and thus leverage herself up to a full understanding.

The upshot of this discussion is that Quine does have something close to an account of what understanding a sentence amounts to for observation sentences. This account, moreover, is an integral part of a central Quinean project, namely making it plausible that the acquisition of cognitive language can be explained in terms which are, by Quine's standards, purely naturalistic.

For non-observation sentences matters are far more complicated. For observation sentences, what must be acquired can be specified in a relatively straightforward manner (although the point made in the last few paragraphs indicates that this is not quite the whole story). For sentences in general, however, there is no specifying the disposition which must be acquired. Quine puts forward an account of how mastery of various idioms involved in some non-observation sentences is or might be acquired, but this account does little to tell us what the understanding of an individual sentence consists in. In his view this is inevitable: for sentences in general, there simply is no account of what is required for understanding.

One way to think about this point is in terms of holism. For an observation sentence, we can say (again with some qualification) that having a disposition to accept it or reject it under the relevant circumstances is what makes one a competent user of the sentence. But nothing of the sort holds for sentences in general. Consider even a relatively simple sentence such as 'The European economy is improving'. What patterns of neural intake must dispose me to accept this sentence, if I am to count as a competent user of it? And what patterns must dispose me to dissent from it? The questions are hopeless. The sight of a newspaper headline containing those words may dispose me to accept it, and surely many who do accept it have little more to go on. Yet the sight of the headline may have no such effect, if I have views of my own, or simply distrust the source. As Quine says about an even simpler example, the theory which connects observation to a verdict on a given sentence 'is composed of sentences associated with one another in multifarious ways, not easily reconstructed even in conjecture' (Quine, 1960: 11).

What Quine says about the way in which language is acquired leaves room for this complexity. On his account, non-observational language is not in general learned by learning that a certain sentence is equivalent to (say) a disjunction of certain observation sentences, or anything of that sort. If it were, then understanding the non-observation sentence would simply be a matter of acquiring a disjunctive disposition, so to speak, a complex but tractable matter. But not so. A sentence learned in one context in which it is equivalent to something already learned, perhaps on the basis of observations, may then be extended to another context, in which the equivalence fails. Quine suggests that something of this sort may hold for relative clauses, which in his account are a crucial step on the way towards acquiring the capacity to refer to objects (Quine, 1974: 93–95). Quite generally, he says, the learner's progress in acquiring language 'is not a continuous derivation, which, if followed backwards, would enable us to reduce scientific theory to sheer observation. It is a progress by short leaps of analogy' (Quine, 1975: 78–79). It is this process that leaves us with the 'multifarious' connections between observation and theory, and means that for most non-observation sentences the question: 'In what does the understanding of this sentence consist?' will simply have no clear answer. It is, indeed, for this reason that Quine would not accept 'understands' as a term of regimented theory: it allows for the formation of questions without clear answers.

In spite of these limitations, Quine's project of accounting for language acquisition does as much as can be done, according to a Quinean view, to account for the understanding of language. It performs one of the functions carried out by Russell's LPL, to the extent that Quine thinks that that can be done. If we could actually formulate the Russellian LPL of a given person, it would show us exactly what goes into that person's understanding of any given sentence, and exactly what that sentence means for the given person. Quine's project cannot do that, but this, in his view, is not a defect. As we have seen above, his view of the basis of language and of the means by which language is acquired sets out to be thoroughly scientific. It leaves no room for anything like Russell's notion of acquaintance. It also gives Quine theoretical reasons for thinking that it is not feasible to say exactly what goes into the understanding of theoretical sentences, and that the picture which suggests that it must be possible is deeply misleading. But to the extent that Quine holds that an account of linguistic understanding *is* available, his account of language acquisition would, if fully carried out, provide it.

4 The two Quinean enterprises

In the first section, I distinguished two aspects of Russell's LPL, the ontological aspect and the aspect having to do with linguistic understanding. The discussions of Quine in the previous two sections show that each of these aspects corresponds to a Quinean project. For Quine, however, they are distinct projects. Here I must address one potential misunderstanding. One might think that the Quinean enterprises are distinct because they concern different languages. Quine's ontological project relies on canonical notation, whereas his concern with the acquisition of language is, at least in the first instance, a concern with the acquisition of ordinary language. But the difference that I am discussing is independent of that point, and would hold even if Quine took ordinary language to be ontologically significant. In that case, there would be a single language under discussion, but still the use of that language to assess the ontology of our theory would be a distinct enterprise from the task of explaining how that language is acquired, and the two enterprises would emphasize different aspects of the language.

Having, I hope, set aside a mistaken view of the way in which the two Quinean enterprises differ, let me try to put forward what I take to be the correct view. Again, it will be useful to allude to Russell, by way of contrast. For Russell, the two enterprises come together. Every meaningful sentence of ordinary language has a fully analyzed version, which is a sentence of the LPL. The structure of that sentence of the LPL *both* shows the ontological commitments of the original sentence *and* shows how one understands it. (In virtue of this second point, the original sentence and the corresponding sentence of the LPL make exactly the same claim: they are synonymous in a very strict sense.) In order to be capable of understanding the sentence, one must understand the simple terms of the corresponding sentence of the LPL, and in order to do *that* one must be acquainted with the entities to which those terms refer. Those same entities also make up the fact which makes the sentence true, if it is true. Our being acquainted with certain entities makes it possible for us to refer to them, and reference is the fundamental relation between language or thought on the one hand and the world on the other hand.

Nothing of the sort can be said for Quine. Not every meaningful sentence is sufficiently objective to have a version in regimented theory, and for those which do there is no claim of synonymy between the ordinary language sentence and the regimented version. His account of our understanding of language – and thus also of the meaningfulness of sentences – depends, at the fundamental level, on the fact that certain patterns of neural intake make us more or less likely to utter or to assent to certain sentences. But that doesn't mean that the sentence or any part of it *refers to* or is in any sense *about* patterns of neural intake. Of course some sentences *are* about patterns of neural intake, but those are theoretical sentences, chiefly of neurophysiology (or perhaps of Quinean philosophy), not observation sentences. Observation sentences are not about the patterns of neural intake to which they are linked. The one-word sentence 'Doggie!', if it's about anything, is presumably about the dog which gives rise to the neural intake that prompts the utterance. From a Quinean point of view, the fundamental relation between language and the world which is relevant to how we understand language, is not aboutness or reference at all. This brings out the difference between Russell and Quine, but it also brings out the difference between the two projects, from Quine's point of view.

Coming at the difference from the other side: The ontology of regimented theory is made up of physical objects and sets. Here the idea of reference is central: the ontology of a theory, on Quine's account, simply is the set of entities which its quantifiers must range over to make the sentences of the theory true. But reference here is not an epistemological relation. Our understanding the sentences of regimented theory does not require any unmediated or pre-theoretical epistemological relation to physical objects or to sets. There is nothing in Quine's thought that is comparable to Russell's relation of acquaintance. We do not have a direct epistemological relation even to observable physical objects, what J.L. Austin called 'moderate-sized specimens of dry goods' [Austin, 1962: 8]. To the contrary: our epistemological relation to those entities is mediated by the theory itself. (It is for this reason that Quine says that he sees 'all objects as theoretical' [Quine, 1981: 20], and for this reason too that the indeterminacy of reference is a coherent doctrine for Quine, as it would not be for Russell.) The sentences of canonical notation refer to sets and to physical objects but, to repeat, reference for Quine is not an epistemological relation, and plays no role in the way we understand those sentences.

So the two Quinean enterprises are quite distinct. That does not, of course, mean that there is no connection between them. The connection is essentially what Quine, in 'Epistemology Naturalized', called 'reciprocal containment, though containment in different senses' (Quine, 1969: 83). Quine is speaking there of the relation between epistemology and natural science. Epistemology, on his account, is a branch of psychology, and hence of natural science. So the containment of epistemology within natural science is quite straightforward. The reverse containment also holds, although in a different way: epistemology is the study of how we come by the knowledge that we have, and our natural science is that knowledge. The same relation holds between the two Quinean enterprises that I have distinguished. We all learn the ordinary language spoken around us; on the basis of that initial leaning, the fortunate among us acquire at least the rudiments of canonical notation. So Quine's account of how language is acquired must, in principle, extend as far as the acquisition of the language of regimented theory, which determines our ontology. If the project of giving such an account were fully realized, the result would be an account in regimented theory of how our language and knowledge – including our knowledge of regimented theory itself – might be acquired. And Quinean ontology lays down the constraints within which the acquisition of language is to be explained – not in terms of our being in direct epistemic relations to propositions, say, but rather in austerely naturalistic terms. So the two projects are related by the same relation of reciprocal containment that Quine discusses in 'Epistemology Naturalized'.

Neither project is fully independent of the other.[8] The project of accounting for the acquisition of language is a project of accounting for it in purely naturalistic terms, i.e. within the constraints laid down by Quinean ontology. Quine's approach to ontology, via regimented theory, presupposes that that theory is adequate to give an account of the world – and, in particular, an account of how language is acquired. These connections, however, do not undermine the idea that the two projects are indeed separable aspects of Quine's overall philosophy.

Let me finally, and very briefly, try to put this discussion in a broader context. Each of the two Quinean enterprises I have discussed has some claim to be called 'Quine's philosophy of language', yet they are quite distinct. This might seem puzzling, but should not. Language is an almost all-pervasive part of human life. There is no reason to expect that it should raise only one kind of philosophical question, or that the various questions that it raises can all be answered by a single, unified account. The two Quinean enterprises emphasize different aspects of the philosophical interest of language, one from the point of view of ontology and the other from the point of view of understanding. The fact that language has these two aspects is surely one of the reasons for its enduring philosophical interest. It faces outward, towards the world that it is about, and inward, towards the mind that understands it. Our assertions, after all, are presumably made true or false by the way the world is, and when they are true they tell us something about the world. Yet they are also what we understand. Russell's LPL provides a unified treatment of these two aspects. Anyone who holds, say, that the meaning of a (one-place) predicate is the property for which it stands, and that understanding the predicate involves an epistemic relation to that property, is also presumably committed to a unified treatment of this kind, at least for predicates. Quine, by contrast, treats the two issues quite separately, and the result is two distinct enterprises, each of which has some claim be thought of as Quine's philosophy of language.

Notes

1. In the fifth of the lectures on 'The Philosophy of Logical Atomism', Russell says 'I do not profess to know what the right analysis of general facts is' (Russell, 1986: 207).
2. In the *Lectures on the Philosophy of Logical Atomism*, Russell uses the term 'logical fiction', but not the term 'logical construction'. In earlier work, he tends to use the latter term, most notably in the 1914 essay 'The Relation of Sense-Data to Physics', where he says: 'The supreme maxim in scientific philosophizing is this: "*Wherever possible, logical constructions are to be substituted for inferred entities.*"' (Russell, 1918 [1914]: 155, emphasis in the original). This is only a terminological change. The reason for it is perhaps that to say that so-and-so is a logical construction suggests that there is such a thing as so-and-so, whereas calling something a logical fiction is less likely to carry that suggestion.
3. Here, and in the rest of my discussion of Russell, I attribute to him a conception of analysis which is, I think, his predominant conception until the spring or summer of 1918. According to that conception, every meaningful sentence expresses a definite thought; vagueness arises because it is often unclear, prior to analysis, which thought is expressed. There are, however, passages in which Russell's work, especially when he is discussing technical issues, which suggest a different view. According to this different view, some sentences do not express any precise or definite thought. Analysis, on this latter view, is pragmatic, and aims at imposing a useful meaning on the sentence being analysed, rather than on uncovering its pre-existing meaning.

There is undoubtedly a tension in Russell's thought here: some passages certainly suggest one view, other passages the other view. My primary purpose in this essay is to interpret Quine; my discussion of Russell is only to supply a clarifying contrast. Nevertheless: I think the conception I attribute to Russell represents the stronger strain in his thought during the period which is my concern. It goes along with the account of understanding which I articulate in the remainder of this section. The other conception, by contrast, leaves him unable to account for understanding in a way compatible with his other commitments. This situation changes rapidly, however, beginning in the spring of 1918, when he drops some of his earlier commitments and begins to articulate the more behaviouristic view of understanding which is set out in *The Analysis of Mind* (Russell, 1921).

For a detailed and insightful discussion of this issue, see James Levine, 'From Moore to Peano to Watson: The Mathematical Roots of Russell's Naturalism and Behaviorism' (Levine, 2009). I am grateful to Levine for correspondence on this matter, and also for letting me read unpublished work on the topic; and also to Fraser McBride, who raised this issue in a question at the conference in Glasgow.

4. The exception is the period 1902–1905, during which Russell held the theory of denoting concepts. According to that theory, understanding a sentence of the form 'the F is G' requires that one stand in an epistemological relation to the denoting concept, *the F*. The fact which makes the sentence true (if it is true), however, does not contain that denoting concept; rather, it contains the denoted object. According to the 1905 theory of descriptions, by contrast, to understand the sentence I must be acquainted with the universals, *F* and *G*, and the relevant logical entities. Those are also the entities which are the constituents of the fact which most directly makes the sentence true. (If the sentence is true then there is also a fact containing the relevant object and the universal G, but that is a distinct fact, and not what is in the most direct sense asserted by the sentence.)
5. In this paragraph and the next, and in a number of other places, I am indebted to Gary Kemp.
6. For related discussion, see (Hylton, 2014), especially section 4; for an articulation of the general view of Quine which lies behind the more specific remarks here and elsewhere, see (Hylton, 2007).
7. Here I rely heavily on the interpretation of Russell given in note 3, above. If one adopts the alternative interpretation of Russell indicated there, his view appears far more Quinean, as Levine emphasizes.
8. I am grateful to Andrew Lugg for pressing me to get clearer on this point and for other comments on earlier versions of this paper.

Bibliography

Austin, J.L. (1962) *Sense and Sensibilia* (Oxford: Oxford University Press).

Hylton, P.W.(2007) *Quine* (London & New York: Routledge).

Hylton, P.W. (2014) 'Significance in Quine', *Grazer Philosophische Studien*, 89: 109–129.

Levine, J. (2009) 'From Moore to Peano to Watson: The Mathematical Roots of Russell's Naturalism and Behaviorism'. In J. Skilters (ed.) 2009, *The Baltic International Yearbook of Cognition, Logic and Communication* 4. Riga: University of Latvia, pp. 1–126.

Quine, W.V. (1960) *Word and Object* (Cambridge Mass : MIT).

Quine, W.V. (1969) *Ontological Relativity and Other Essays* (New York: Columbia University Press).

Quine, W.V. (1974) *Roots of Reference*, (La Salle, IL: Open Court).

Quine, W.V. (1975a) 'Mind and Verbal Dispositions'. In S. Guttenplan (ed.) 1975, *Mind and Language*. Oxford: Clarendon Press, pp. 83–95.

Quine, W.V. (1975b) 'The Nature of Natural Knowledge'. In S. Guttenplan (ed.) 1975, *Mind and Language*. Oxford: Clarendon Press pp. 76–81.

Quine, W.V. (1976 [1957]) 'The Scope and Language of Science', *British Journal for Philosophy of Science* 8: 1–17; reprinted in *Ways of Paradox* (Cambridge Mass: Harvard University Press; expanded edition 1976), pp. 228–245.

Quine, W.V. (1981) *Theories and Things* (Cambridge Mass: Harvard University Press).

Russell, B. (1918 [1914]) 'The Relation of Sense-Data to Physics' *Scientia*, 16: 1–27; reprinted in *Mysticism and Logic and Other Essays*. New York, London: Longmans, Green & Co., pp. 145–179.

Russell, B. (1986 [1918]) 'The Philosophy of Logical Atomism' . In J.G. Slater (ed.) 1986, *The Collected Papers of Bertrand Russell*. London: George Allen and Unwin, pp. 157–244.

Russell, B. (1999 [1912]) *Problems of Philosophy* (Minneola NY: Dover).

Russell, B. (1921) *The Analysis of Mind* (London: George Allen & Unwin).

10
Reading Quine's Claim That No Statement Is Immune to Revision

Gary Ebbs

Most critics and defenders of Quine's arguments in 'Two Dogmas of Empiricism' (Quine, 1953a) have read his claim that 'no statement is immune to revision' as the claim that *for every statement S that we now accept there is a possible rational change in beliefs that would lead one to reject S*. For reasons I'll explain below, I paraphrase this latter claim as follows:

(R)   For every sentence S that a subject A accepts at a time t_1, there is a possible rational revision of the beliefs A holds at t_1 that (i) leads A, or another subject B, rationally to judge, at some later time t_2, that S is false, and (ii) allows for a homophonic translation of S, as A uses it at t_1, by S, as A or B uses it at t_2.

This standard reading of Quine's claim that 'no statement is immune to revision' faces two serious problems. First, in 'Two Dogmas' Quine does not even mention the issues about diachronic changes in belief or homophonic translation that are relevant to supporting (R). He writes as if he thinks he has a simple, direct argument that does not rest on such considerations. Second, Quine's own views about translation lead him to conclude, apparently contrary to (R), that some revisions in our current beliefs *would* alter the meanings of some of our words to the point that we would not accept a homophonic translation of some of the sentences we held true before the revision.

Until recently I accepted the standard reading despite these problems. On the reading I now prefer, however, Quine's claim is that no statement is immune to retraction,[1] understood as follows:

(P)   No statement we now accept is guaranteed to be part of every scientific theory that we will later come to accept.

123

I will argue that in paragraph two of section 6 Quine combines this uncontroversial observation with arguments and proposals from previous parts of 'Two Dogmas' to support his conclusion that 'it [is] folly to seek a boundary between synthetic statements, which hold contingently on experience, and analytic statements, which hold come what may' (Quine, 1953a: 43). The key to my alternative interpretation is to see that in paragraph one of section 6 Quine sketches a bold new naturalistic explication of the traditional notion of empirical confirmation, and that his aim in paragraph two is to show that the explication of confirmation he sketches in paragraph one is of no help in characterizing a boundary between analytic and synthetic statements. This is the last of a series of clarifications and observations in 'Two Dogmas' that in Quine's view together show that it is folly to seek a boundary between analytic and synthetic statements, or, in other words, that we have no reason to suppose that such a distinction can be drawn in a language suited for and used in the mature natural sciences.

1   A first look at the context of Quine's claim that 'No statement is immune to revision'

Quine's claim that 'No statement is immune to revision' appears in paragraph two of section 6 (the final section) of 'Two Dogmas of Empiricism'. By the start of section 6, Quine takes himself to have shown in the previous five sections of the paper that there is no way to draw an analytic-synthetic boundary in terms of an extensional specification of logical truth supplemented by definitions (sections 1–2); an extensional specification of logical truth supplemented by a synonymy relation defined in terms of substitutivity *salve veritate* (section 3); semantic rules laid down for an artificial language (section 4); or confirmation by experience (section 5).

In the first paragraph of section 6, Quine likens science to 'a field of force whose boundary conditions are experience', and sketches some consequences of this comparison, including the consequence that 'No particular experiences are linked with any particular statements in the interior of the field, except indirectly through considerations of equilibrium affecting the field as a whole'. In paragraph two he writes,

> If this view is right, it is misleading to speak of the empirical content of an individual statement – especially if it is a statement at all remote from the experiential periphery of the field. Furthermore it becomes folly to seek a boundary between synthetic statements, which hold

contingently on experience, and analytic statements, which hold come what may. Any statement can be held true come what may, if we make drastic enough adjustments elsewhere in the system. Even a statement very close to the periphery can be held true in the face of recalcitrant experience by pleading hallucination or by amending certain statements of the kind called logical laws. Conversely, by the same token, no statement is immune to revision. Revision even of the logical law of the excluded middle has been proposed as a means of simplifying quantum mechanics; and what difference is there in principle between such a shift and the shift whereby Kepler superseded Ptolemy, or Einstein Newton, or Darwin Aristotle? (Quine, 1953a: 43)

The paragraph begins with a conditional whose antecedent is 'this view is right' and whose consequent is a conjunction of 'it is misleading to speak of the empirical content of an individual statement' and – as signaled by the word 'Furthermore' – 'it becomes folly to seek a boundary between synthetic statements, which hold contingently on experience, and analytic statements, which hold come what may'. In the rest of the paragraph, Quine presents an argument in support of the conditional 'if this view is right, then it becomes folly to seek a boundary between synthetic statements, which hold contingently on experience, and analytic statements, which hold come what may'. The main premises of Quine's argument for this conditional are 'any statement can be held true come what may' and 'no statement is immune to revision'.

2 The standard interpretation

In their influential paper, 'In Defense of a Dogma', H.P. Grice and P.F. Strawson say that Quine takes his assertion that no statement is immune to revision 'to be incompatible with acceptance of the distinction between analytic and synthetic statements' (Grice and Strawson, 1956: 154). They try to discredit Quine's argument in paragraph two of section 6 by pointing out that there is a perfectly ordinary interpretation of 'no statement is immune to revision' on which this sentence is 'not incompatible with acceptance of the distinction, but is, on the contrary, most intelligibly interpreted in a way quite consistent with it' (Grice and Strawson, 1956: 154). They reason as follows:

Any form of words at one time held to express something true may, no doubt, at another time, come to be held to express something false.... Where such a shift in the sense of the words is a necessary condition of the change in truth-value, then the adherent of the distinction will say that the form of words in question changes from expressing an analytic statement to expressing a synthetic statement.... If we can make sense of the idea that the same form of words, taken in one way (or bearing one sense), may express something true, and taken another way (or bearing another sense), may express something false, then we can make sense of the idea of conceptual revision. And if we can make sense of this idea, then we can perfectly well preserve the distinction between the analytic and the synthetic, while conceding to Quine the revisability-in-principle of everything we say. (Grice and Strawson, 1956: 157)

This challenge to Quine's reasoning has put Quineans on the defensive. In response to it, Hilary Putnam (in Putnam, 1962 and many other papers, including Putnam, 1979) argued, in effect, that even if we cannot *now* see how we could judge that a sentence *S* is false without changing its meaning to the point where we would no longer interpret it homophonically into our new body of beliefs, we may find *later* that we *can* judge that *S* is false while still translating our previous uses of *S* into our (revised) theory homophonically.[2]

Putnam's influential response to the Grice-Strawson criticism led many interpreters (including me) to read Quine's claim that 'No statement is immune to revision' as the claim that no statement is immune to *rejection*, where the notion of rejection is understood partly in terms of homophonic translation. As Gilbert Harman explains, on Quine's view,

There is no sharp, principled distinction between changing what one means and changing what one believes. We can, to be sure, consider how to translate between someone's language before and after a given change in view. If the best translation is the homophonic translation, we say there has been a change in doctrine; if some other (nonhomophonic) translation is better, we say there has been a change in meaning. (Harman, 1994: 141)

These Quinean points about meaning and homophonic translation are integral to the standard interpretation of Quine's claim that no statement is immune to revision. According to Cory Juhl and Eric Loomis, for instance,

Quineans...introduce a surrogate for "means the same," when they appeal to the notion of a "good" or "best" translation scheme.

A sentence "retains its meaning" across changes, on this Quinean picture, just in case the sentence would or should be translated homophonically across the change in language. But given this "surrogate" for synonymy, it seems as if Quineans can now make sense of what Carnap, Grice and Strawson, and a host of others are worried about when considering the possibility of "giving up" a statement on the basis of empirical evidence. *It is not enough for the Quinean to show that we could give up our practice of asserting sentence s, for some purportedly analytic s.* Rather, in order to address the worries of his opponents, the Quinean must show that the sentence can be given up while retaining its meaning across the change in language, that is, he must show that the sentence is such that it could be translated homophonically across the change. *It may be that some sentences might plausibly stop being asserted, but homophonic translatability imposes a further constraint, and narrows the range of sentences which meet it. Whether any of the usual examples (bachelorhood in the face of new marriage laws, etc.) meet this constraint is likely to remain controversial.* (Juhl and Loomis, 2010: 116, my emphases)

On this standard interpretation, to discredit the assumption that there is a boundary between analytic and synthetic statements, as Juhl and Loomis say, 'It is not enough for the Quinean to show that we could give up our practice of asserting sentence s, for some purportedly analytic s. Rather, in order to address the worries of his opponents, the Quinean must show that the sentence can be given up while retaining its meaning across the change in language'. On this interpretation, if Quine's claim that 'no statement is immune to revision' is to be understood in a way that engages with the worries of his opponents, it must be understood as follows:

(R)   For every sentence S that a subject A accepts at a time t_1, there is a possible rational revision of the beliefs A holds at t_1 that (i) leads A, or another subject B, rationally to judge, at some later time t_2, that S is false, and (ii) allows for a homophonic translation of S, as A uses it at t_1, by S, as A or B uses it at t_2.

Following Grice and Strawson, critics of Quine's reasoning in paragraph two of section 6 of 'Two Dogmas, say, in effect, that in this paragraph Quine does not make a convincing (or indeed, any) case for (R). Defenders of Quine's reasoning in paragraph two have for the most part (tacitly) accepted this criticism. Following Hilary Putnam, Quineans typically respond to the criticism by trying to provide support for (R). It is now common for both Quineans and critics of Quine to read (R) back into paragraph two of section 6, and to conclude that the argument there is at best incomplete.

3   Two problems for the standard reading

There are two serious problems for the standard reading of Quine's reasoning in paragraph 2 of section 6 of 'Two Dogmas'.

First, Quine's brief and sketchy presentation of his reasoning strongly suggests that he thought a reader who adopts his 'field of force' description of science would not need much detail to be convinced of his claim that 'no statement is immune to revision' and to agree that it implies that there are no analytic statements. If the claim amounts to (R), however, it is neither obvious nor uncontroversial. To support (R) one would need to show that radical changes in belief never bring about radical changes in meaning of the sort that would lead one to reject a homophonic translation of a previously uttered sentence into one's new theory. But Quine does not say anything about these difficult issues in the second paragraph of section 6 (or anywhere else in 'Two Dogmas'). He writes as if he thinks he has a simple, direct argument that does not rest on such considerations.

Second, in the Introduction to the first edition of *Methods of Logic*, published in 1950 – the same year in which Quine first presented 'Two Dogmas of Empiricism' at the Eastern meeting of the APA in Toronto and the year *before* that paper was published in *The Philosophical Review* – Quine writes that mathematics and logic are 'central to our conceptual scheme' in the sense that they 'can easily be held immune to revision on principle', and therefore 'tend to be accorded such immunity, in view of our conservative preference for revisions which disturb the system least' (Quine. 1950: xiii). He adds that 'it is perhaps the same to say, as one often does, that the laws of mathematics and logic are true simply by virtue of our conceptual scheme' (Quine, 1950: xiv), and then notes that

> It is also often said that the laws of mathematics and logic are true by virtue of the meanings of the words '+', '=', 'if', 'and', etc., which they contain. This I can also accept, for I expect it differs only in wording from saying that the laws are true by virtue of our conceptual scheme. (Quine, 1950: xiv)

He also observes that

> Mathematical and logical laws themselves are not immune to revision if it is found that essential simplifications of our whole conceptual scheme will ensue. There have been suggestions, stimulated largely by quandaries of modern physics, that we revise the true-false dichotomy of current logic in favor of some sort of tri- or *n*-chotomy. (Quine, 1950: xiv)

The point Quine makes in the second sentence of this passage is clearly similar to his observation in the second paragraph of section 6 of 'Two Dogmas of Empiricism' that 'Revision even of the logical law of the excluded middle has been proposed as a means of simplifying quantum mechanics'. But the crucial passage from Quine's Introduction to the first edition of *Methods of Logic* – the passage that clearly conflicts with attributing (R) to Quine – is the following:

> Thus the laws of mathematics and logic may, despite all "necessity", be abrogated. But this is not to deny that such laws are true by virtue of the conceptual scheme, or by virtue of meanings. *Because these laws are so central, any revision of them is felt to be the adoption of a new conceptual scheme, the imposition of new meanings on old words.* (Quine, 1950: xiv, emphasis added)

One might suspect that these passages were written before Quine finally made up his mind, while writing 'Two Dogmas of Empiricism', to give up the analytic-synthetic distinction, and that Quine just did not get around to revising the passages before the first edition of *Methods of Logic* went to press. In fact, however, the crucial italicized sentence in the passage just quoted also appears in all *subsequent* editions of *Methods of Logic* – it appears on p. xiv of the 1959 revised edition, and on p. 3 of both the third (1972) and fourth (1982) editions. Moreover, in 'Carnap and Logical Truth', written in 1954, three years after 'Two Dogmas' was first published, Quine argues that we should question our translation of an apparently sincere utterance if under the translation, the utterance counts as expressing the negation of an obvious logical law. As Quine later stressed, his view of translation, as applied to logical statements of a person's theory, implies that a logician who wishes to revise an established logical law faces a 'predicament: when he tries to deny the doctrine, he only changes the subject' (Quine, 1986: 81). As the above passages from the Introduction to the 1950 edition of *Methods of Logic* show, this was clearly his view already at the time he wrote 'Two Dogmas of Empiricism'. He was therefore not in a position to affirm (R) for any sentence that he used to express one of these obvious logical laws. Yet, as we have seen, in paragraph two of section 6, in support of his claim that no statement is immune from revision, he writes, 'Revision even of the logical law of the excluded middle has been proposed as a means of simplifying quantum mechanics' (Quine, 1953a: 43). This strongly suggests that Quine's claim that no statement is immune to revision should not be interpreted as (R).

Despite these problems, until recently I accepted the standard interpretation of Quine's reasoning in paragraph two of section 6. Like many others, I assumed that Putnam-style counter-examples to analyticity – counterexamples that support, even if they do not conclusively establish, (R) – provide the best grounds for Quine's claim no statement is immune to revision. Thus bewitched, I did not look carefully for a plausible alternative to the standard interpretation, and, of course, did not find one, either. While recently trying to explain to myself (and my students) exactly what is going on in paragraph two of section 6, however, I studied that paragraph more closely, and discovered a better interpretation, one that avoids the problems just explained.

4   An alternative interpretation

On the alternative interpretation that I now prefer, Quine's reasoning in paragraph two of section 6 is the last of a series of clarifications and observations in 'Two Dogmas' that in Quine's view together show that it is fruitless to seek a boundary between analytic and synthetic statements. Quine does not claim that there is no such boundary. His conclusion in paragraph two is that 'it becomes folly to seek a boundary between synthetic statements, which hold contingently on experience, and analytic statements, which hold come what may'. He is similarly subtle in section 4, where he announces that

> For all its a priori reasonableness, a boundary between analytic and synthetic statements simply has not been drawn. That there is such a distinction to be drawn at all is an unempirical dogma of empiricists, a metaphysical article of faith. (Quine, 1953a: 37)

These words signal that Quine is employing Carnap's own preferred method for revealing the emptiness of a philosopher's words – a method partly inspired by Wittgenstein's pronouncement, in his *Tractatus-Logio-Philosophicus*, that

> the correct method in philosophy [is] to say nothing except what can be said, i.e. propositions of natural science – i.e. something that has nothing to do with philosophy – and then, whenever someone else wanted to say something metaphysical, to demonstrate to him that he had failed to give a meaning to certain signs in his propositions. (Wittgenstein, 1961 [1921]: 73–74)

As I read it, Carnap's and Quine's dispute about the analytic-synthetic distinction is at root a dispute about whether such a distinction can be drawn in a language suited for and used in the mature natural sciences.

Quine's strategy in 'Two Dogmas' is to demonstrate to Carnap and others that none of the minimally plausible strategies for giving clear meanings to the terms 'analytic' and 'synthetic' is successful, so the terms should be dropped from a properly scientific philosophy. To pursue this strategy, at each new step in his reasoning, including the step he takes in paragraph two of section 6, Quine needs to rely on clarifications and observations that he makes in earlier parts of the paper. As signaled by its first sentence, which begins with the words 'If this view is right', Quine's reasoning in paragraph two relies immediately and obviously on the view of science that he briefly sketches in the preceding paragraph, which I quote here in full:

> The totality of our so-called knowledge or beliefs, from the most casual matters of geography and history to the profoundest laws of atomic physics or even of pure mathematics and logic, is a man-made fabric which impinges on experience only along the edges. Or, to change the figure, total science is like a field of force whose boundary conditions are experience. A conflict with experience at the periphery occasions readjustments in the interior of the field. Truth values have to be redistributed over some of our statements. Reevaluation of some statements entails reevaluation of others, because of their logical interconnections – the logical laws being in turn simply certain further statements of the system, certain further elements of the field. Having reevaluated one statement we must reevaluate some others, which may be statements logically connected with the first or may be the statements of logical connections themselves. But the total field is so underdetermined by its boundary conditions, experience, that there is much latitude of choice as to what statements to reevaluate in

the light of any single contrary experience. No particular experiences are linked with any particular statements in the interior of the field, except indirectly through considerations of equilibrium affecting the field as a whole. (Quine, 1953a: 42–43)

In this paragraph Quine writes of 'readjustments' occasioned by 'a conflict with experience', readjustments that reflect 'reevaluations of some statements'. Since 'there is much latitude of choice as to what statements to reevaluate in the light of any single contrary experience', what holds a statement *S* in place in the 'interior' of the field is not our recognition of a theory-independent standard of confirmation for *S*, but our actual, current *acceptance* of *S* – our evaluation of *S* as true – and the logical and explanatory relations between *S* and other statements, relations that are themselves settled by our acceptance of statements of laws of logic, mathematics, and physics. Particular experiences are 'linked with', and in that sense, *confirm*,[3] particular statements in the interior of the field only 'indirectly through considerations of equilibrium affecting the field as a whole'. In short, on the view Quine sketches in the first paragraph of section 6, an inquirer's acceptance of a statement establishes a relation between sentences and experience that does not exist apart from his or her acceptance of the theory, and to say that a statement is *confirmed* is just to say that one accepts it as part of one's best current theory.

This is a naturalistic explication (or appropriation, if you prefer) of Carnap's radical view that there is no relation of confirmation apart from our *decisions* about how to relate statements to experience. Quine's bold step is to do without Carnap's idea that some statements of a theory are analytic (i.e. true in virtue of semantical rules that we have laid down for the language of the theory) and others are synthetic, (i.e. have truth values that are not settled by semantical rules that we have laid down for the language of the theory). For reasons that Quine explains in section 4 of 'Two Dogmas', he thinks Carnap fails to provide a satisfactory explication of this idea. Since all of Carnap's attempts to clarify the idea that some statements have empirical content presuppose his explications of analyticity in terms of semantical rules, once Quine takes himself to have shown that the idea of analyticity cannot be explained in terms of semantical rules, he can easily show in section 5, drawing on observations about the holism of theory testing that Carnap himself had made in previous publications, that a boundary between analytic and synthetic statements cannot be drawn on the basis of relations between sentences and experiences that supposedly confirm them. (Many readers of 'Two Dogmas' do not realize that Quine's point in section 5 is one that Carnap himself would readily accept, namely, that if there is no independent account of the boundary between analytic and synthetic statements, an appeal to relations between sentences and experiences that supposedly confirm them will not help us to draw it.) What is left, according to Quine, as he later explained in *Word and Object*, section 6, is just 'scientific method' which 'produces theory whose connection with all possible surface irritation consists solely in scientific method itself, unsupported by ulterior controls' (Quine, 1960: 23).

In paragraph two Quine explains why one cannot exploit this minimalist explication of confirmation – the only positive bit of theorizing about confirmation that he presents in 'Two Dogmas' – to draw a boundary between analytic and synthetic statements. He reasons as follows. If one grants his first point in the paragraph, namely, that one consequence of his minimalist explication of confirmation is that 'it is misleading to speak of the empirical content of an individual statement – especially if it is a statement at all remote from the experiential periphery of the field', then the most plausible way to exploit his minimalist explication of confirmation to draw a boundary between analytic and synthetic statements, would be to equate 'S is analytic' with a *new* explication of 'S is confirmed come what may' – the same phrase for which in section 5 of 'Two Dogmas' he could find no clarification in terms of a supposedly substantive, theory-independent relation of confirmation by experience. In the context of the minimalist explication of confirmation that Quine sketches in the first paragraph of section 6, to say that a statement is confirmed is just to say that one accepts it as part of one's best current theory. Hence if we adopt Quine's minimalist explication of confirmation, and give up hope of drawing the analytic-synthetic boundary in terms of a supposedly substantive, theory-independent relation of confirmation by experience, as Quine recommend, then 'S confirmed come what may' simply amounts to 'S is a statement of our current theory and S is guaranteed to be part of every theory that we will later come to accept'.

Quine thinks it is worth considering, if only briefly, the question whether a boundary between analytic and synthetic sentences can be drawn by appealing to this minimalist explication of 'S is confirmed come what may'.[4] He points out, in effect, that it is enough to formulate this question clearly to see that the answer to it is 'No'. The problem is that all parties to the dispute about analyticity should accept

(P) No statement we now accept is guaranteed to be part of every scientific theory that we will later come to accept.

from which it follows that no statement is confirmed come what may in the minimalist sense in question.

On this reading, in paragraph two Quine does not assert, conversationally imply, or presuppose (R). Instead, he points out, in effect, that

IF

(a) his arguments in sections 1–4 are successful, and
(b) as he argues in section 5, no boundary between analytic and synthetic sentences can be drawn in terms of confirmation by experience, and
(c) as he suggests in paragraph one of section 6, to say that a statement is confirmed is just to say that one accepts it as part of one's best current theory,

THEN

(d) To say of a statement which we accept now, and is therefore confirmed in Quine's minimalistic sense, that it is 'confirmed come what may' is to say that it is guaranteed to be part of every scientific theory that we will later come to accept.

Quine then notes, in effect, that all parties to the dispute about analyticity should accept (P), from which it follows that no statement S that we now accept is 'analytic' in the proposed minimalistic sense. Since this is the *only* sense of 'analytic' that Quine can think of that has not already been ruled out by the arguments in sections 1–5 of 'Two Dogmas of Empiricism' and that is compatible with the minimalist account of confirmation that he sketches in paragraph one of section 6, he concludes that if the view sketched in the first paragraph of section 6 is right, 'it becomes folly to seek a boundary between synthetic statements, which hold contingently on experience, and analytic statements, which hold come what may'.

5   Why Carnap accepts (P) but rejects Quine's conclusion

Quine had reason to be confident that Carnap, whose efforts to explicate the analytic-synthetic distinction are Quine's main focus in 'Two Dogmas', would readily accept (P). Already in 1937, in *Logical Syntax of Language* (Carnap, 1937), Carnap held that if a sentence of our theory logically contradicts an observation sentence we accept, then 'some change must be made in the system', but *'There are no established rules for the kind of change which must be made'* (Carnap, 1937: 318; emphasis in the original). He emphasizes that all rules, including those that (according to him) settle the logic of the language, and those that concern the laws of one's physical theory, 'are laid down with the reservation that they may be altered as soon as it is expedient to do so' (Carnap, 1937: 318). When paraphrased in the terms of Quine's minimalist account of confirmation, without any reliance on rules or on an analytic-synthetic distinction, Carnap's point becomes Quine's point that 'there is [so] much latitude of choice as to what statements to reevaluate in the light of any single contrary experience' that we may reevaluate even 'the statements of logical connections themselves'.

Carnap reaffirms his commitment to this methodological principle in his 1963 'W. V. Quine on Logical Truth', where he writes (very generously, since, as he surely must have recalled, he made all these points Carnap 1937):

> Quine shows…that a scientist, who discovers a conflict between his observations and his theory and who is therefore compelled to make a readjustment somewhere in the total system of science, has much latitude with respect to the place where a change is to be made. In this procedure, no statement is immune to revision, not even the statements of logic and of mathematics. There are only practical differences, and these are differences in degree, inasmuch as a scientist is usually less willing to abandon a previously accepted general empirical law than a single observation sentence, and still less willing to abandon a law of logic or of mathematics. With all this I am entirely in agreement. (Carnap, 1963: 921)

Thus Carnap interprets Quine's claim that no statement is immune to revision as (P), and accepts it. He apparently takes (P) to be an immediate *consequence* of the 'field of force' picture of science that Quine sketches in the first paragraph of section 6. Carnap therefore sees that Quine's reasoning in paragraph two draws consequences from points Quine makes in the previous paragraph, especially the point that adjustments can be made anywhere in case of a conflict with experience. So far so good.

In responding to Quine's reasoning in paragraph two of section 6, however, Carnap does not acknowledge that in section 4 of 'Two Dogmas', Quine takes himself to have discredited Carnap's efforts to define 'analytic' and 'synthetic' in terms of rules of a language system.[5] Instead, to resist Quine's conclusion, Carnap rehearses his view that whether a sentence is analytic or not in a given language is settled by 'rules of the language', and therefore 'has nothing to do with, transitions 'from a language L_n to a new language L_{n+1}'. He emphasizes that in his view,

> "analytic in L_n" and "analytic in L_{n+1}" are two different concepts. That a certain sentence S is analytic in L_n means only something about the status of S within the language L_n: as has often been said, it means that that truth of S in L_n is based on the meanings in L_n of the terms occurring in S. (Carnap, 1963: 921)

What Carnap apparently does not see is that in the context of Quine's earlier arguments, the first paragraph of section 6 sketches a minimalist account of *confirmation* that does not presuppose or require that a boundary between analytic and synthetic statements can be drawn. This new minimalist account of confirmation provides us with no way to draw a boundary between analytic and synthetic statements by examining a theory at given time, without bringing in considerations about how it may be revised. This leaves us with one remaining proposal to consider and rule out – namely, the proposal that 'S is analytic' means 'S is confirmed come what may', where this amounts to 'S is guaranteed to be part of every theory that we will later come to accept'. Quine's point in paragraph two is that this proposal is immediately undermined by (P), which both he and Carnap accept.

6   Why Grice and Strawson accept (P) but reject Quine's conclusion

Although Quine did not have Grice and Strawson in mind when he wrote 'Two Dogmas', they would also readily accept (P). 'The point of substance (or one of them) that Quine is making, by this emphasis on revisability', they write, 'is that there is no absolute necessity about the adoption or use of any conceptual scheme whatever' (Grice and Strawson, 1956: 157). It is clear from the context that Grice and Strawson intend to use the word 'conceptual scheme' in a way that Quine uses it in 'Two Dogmas', as, for example, when he writes of 'the conceptual scheme of science', on page 44 (of Quine, 1953a).[6] For Quine a conceptual scheme is simply a body of beliefs related to experience in the way he sketches in the first paragraph of section 6. As we saw, revisions of a conceptual scheme are occasioned by re-evaluations of some of its statements, including, in some cases, the retraction of some statements.

Hence if there is no conceptual scheme that every reasonable inquirer must adopt, as Grice and Strawson grant, then (P) is true.

Grice and Strawson assume that Quine's reasoning in paragraph two is supposed to be independent of his reasoning in earlier parts of 'Two Dogmas'. This assumption leads them to miss Quine's strategy of exposing the fruitlessness of the most plausible ways of clarifying 'analytic' and 'synthetic' and thereby showing that those who use these words have not given them any meaning. Grice and Strawson therefore search for an interpretation of Quine's claim that no statement is immune to revision according to which the claim by itself implies that there are no analytic statements.[7] In effect, they believe, on the most plausible interpretation, 'no statement is immune to revision' amounts to (R). Taking this interpretation for granted, they object that Quine has not established (R), and that all he is clearly entitled to is (P), which by itself does not imply that there are no analytic statements. It should now be clear that this objection is irrelevant to Quine's argument in paragraph two of section 6.

Grice and Strawson have another strategy for defending analyticity, however: they invite us to consider examples of statements sincere affirmations of which we fail to understand.[8] They invite us to compare, for instance,

(1) My neighbor's three-year-old child understands Russell's theory of types.

with

(2) My neighbor's three-year-old child is an adult.

Grice and Strawson say that if person X sincerely utters (1), we are likely to ask X for proof of X's unlikely claim, whereas, if person Y sincerely utters (2), 'we would be inclined to say we just don't understand what Y is saying, and to suspect that he just does not know the meaning of some of the words he is using' (Grice and Strawson, 1956: 151). They take this sort of example to provide an informal explanation of the commonsense notion of analyticity.[9] For these types of explanation, 'The distinction on which we ultimately come to rest is that between not believing something and not understanding something' (Grice and Strawson, 1956: 151).

Grice and Strawson (and many others who follow them) assume, in effect, that if we cannot *make sense* of rejecting (asserting the negation of) a sentence *S*, such as 'Bachelors are unmarried' or 'If time flies then time flies', then *S* expresses an analytic truth – i.e. *S* is 'true come what may' in a sense that contradicts (R). In effect, they take our failure to make sense rejecting a sentence *S* that we now accept as grounds for the following claim:

> (A)   For at least one sentence *S* that we now accept there is no possible rational revision of the beliefs we now hold that (i) leads us, or other subject, *B*, rationally to judge, at a later time *t*, that *S* is false, and (ii) allows for a homophonic translation of *S*, as we use it now, by *S*, as we or *B*, use it at time *t*, after the revision.

Let us call this line of reasoning – from our failure to make sense of rejecting *S* to the conclusion that *S* is analytic, in the sense specified by (A) – the *argument from incomprehension*. This argument supports the standard reading of 'no statement is immune from revision', as (R). For it suggests that to challenge (A), one would have to provide grounds for (R), which is logically incompatible with (A).

In *Word and Object*, Quine responds to Grice and Strawson's argument from incomprehension, indirectly, as follows:

> Sentences like "No unmarried man is married", "No bachelor is married", and "2+2 = 4" have a feel that everyone appreciates. Moreover the notion of "assent come what may" gives no fair hint of the intuition involved. One's reaction to denials of sentences typically felt as analytic has more in it of one's reaction to ungrasped foreign sentences. Where the sentence concerned is a law of logic … dropping [it] disrupts a pattern on which the communicative use of a logical particle heavily depends. Much the same applies to "2+2 = 4", and even to "The parts of the parts of the thing are parts of the thing". The key words here have countless further contexts to anchor their usage, but somehow we feel that if our interlocutor will not agree with us on these platitudes there is no depending on him in most of the other contexts containing the terms in question. (Quine, 1960: 66–67)

This passage develops Quine's observation in his 1950 Introduction to *Methods of Logic* that '[b]ecause th[e] laws [of logic and mathematics] are so central, any revision of them is felt to be the adoption of a new conceptual scheme, the imposition of new meanings on old words' (Quine, 1950: xiv). Now, in *Word and Object*, however, he compares

'One's reaction to denials of sentences typically felt as analytic' with 'one's reaction to ungrasped foreign sentences' and proposes to explain both by reflecting on maxims for translation. The basic idea is that translation proceeds in accord with a maxim that Quine states earlier in *Word and Object* as follows:

> The maximum of translation underlying all this is that assertions startlingly false on the face of them are likely to turn on hidden differences of language. This maxim is strong enough in all of us to swerve us even from the homophonic method that is so fundamental to the very acquisition and use of one's mother tongue. (Quine, 1960: 59)

According to Quine, to say 'assertions startlingly false on the face of them are likely to turn on hidden differences of language', is *not* to claim that there are assertions that are analytic in either the confirmational sense of (P) or the semantic sense of (A). Commenting on cases in which one's response to the retraction of a sentence is similar to one's response to an ungrasped foreign sentence, Quine writes, '[Such] intuitions are blameless in their way, but it would be a mistake to look to them for a sweeping epistemological dichotomy between analytic truths as byproducts of language and synthetic truths as reports on the world' (Quine, 1960: 67).

In support of this evaluation, he cites 'Two Dogmas'.[10] Quine would surely not have cited 'Two Dogmas' in this context if he thought that its arguments were vulnerable to the argument from incomprehension. Moreover, Quine rejects the argument from incomprehension for reasons that suggest he should not affirm (R). A central obstacle to affirming (R) is that we know of some statements, including basic laws of logic, our retraction of which would, as Quine says, 'disrupt a pattern on which the communicative use of a logical particle heavily depends'. We may therefore be unable, at present, to understand the assertions of a speaker who retracts such a law. But if we cannot understand the assertions of the speaker, then *a fortiori* we cannot interpret them homophonically into our current theory. Quine's reasons for rejecting the argument from incomprehension therefore also cast doubt on the standard interpretation, according to which he affirms (R).

Quine's rejection of the argument from incomprehension, and his adherence to the above maxim of translation, applies to high-level theorizing in physics, as well. He writes, for instance,

> In the case of wavicles...our coming to understand what the objects are is for the most part just our mastery of what the theory says about

them. We do not learn first what to talk about and then what to say about it. (Quine, 1960: 16)

Quine directly follows these remarks by tracing their consequences for an imagined discussion between two physicists about whether neutrinos have mass. He writes:

> Are they discussing the same objects? They agree that the physical theory which they initially share, the preneutrino theory, needs emendation in light of an experimental result now confronting them. The one physicist is urging an emendation which involves positing a new category of particles, without mass. The other is urging an alternative emendation which involves positing a new category of particles with mass. The fact that both physicists use the word 'neutrino' is not significant. To discern two phases here, the first an agreement as to what the objects are (viz. neutrinos) and the second a disagreement as to how they are (massless or massive), is absurd. (Quine, 1960: 16)

I take these remarks to imply that on Quine's view, just as in the case of a basic logical law, to retract some part of our present theory of, say, neutrinos, would be to disrupt a pattern on which the communicative use of the term 'neutrino' now heavily depends. One might therefore find oneself unable to understand some of the utterances of a physicist who retracts a law of neutrinos. Such a physicist may continue to use the word-form 'neutrino', but, in Quine's view, we should not translate his word-form 'neutrino' homophonically.

Quine later pushed this point about translation even further, noting that 'If the natives are not prepared to assent to a certain sentence in the rain, then equally we have reason not to translate the sentence as "It is raining"' (Quine, 1986: 82).

Unlike logical laws, physical laws and utterances of 'It's raining' are not supposed to be analytic. Hence Quine's account of failures of comprehension in terms of the above-stated maxim of translation applies not only to our so-called intuitions about analyticity, but, more generally, to any case in which '[a] native's unreadiness to assent to a certain sentence gives us reason not to construe the sentence as saying something whose truth should be obvious to the native at the time' (Quine, 1986: 82). Contrary to what Grice and Strawson argue, according to Quine, the phenomenon of incomprehension is not particularly relevant to analyticity, and can be explained by a maxim of translation, without any appeal to the notion of analyticity.[11]

7   Should Quine affirm (R)?

One might still be inclined to reason as follows:

> Since Grice and Strawson in effect affirm (A), to oppose their view Quine would need to provide grounds for rejecting (affirming the negation of) (A). Given that (A) is logically incompatible with (R), the only way to provide such grounds is to provide grounds for affirming (R). Hence to answer the Grice-Strawson challenge, Quine must provide grounds for affirming (R). The standard interpretation is therefore correct, and Quine's reasoning is vulnerable in just the ways that Grice and Strawson pointed out.

This reasoning goes wrong in its assumption that to oppose Grice and Strawson's affirmation of (A), Quine needs to provide grounds for rejecting (A). In fact, to oppose Grice and Strawson's affirmation of (A), it is enough for Quine to have grounds for declining to affirm (A). Given that (A) is logically incompatible with (R), to oppose Grice and Strawson's affirmation of (A), it would be enough for Quine to have grounds for declining to reject (R).

My interpretation of Quine's reasoning in paragraph two of section 6 does not speak directly to the question of what Quine's attitudes toward (A) and (R) should be. As we have seen, however, he has no good reason to affirm (R), given his views of translation. He should therefore *decline to affirm* (R). Should he go so far as to reject (i.e. affirm the negation of) (R)? Since (R) is obviously logically incompatible with (A), to *reject* (R) is thereby also to *affirm* (A). But Quine has no good reason to affirm (A), either. He should therefore *decline to affirm* (A). Given the logical incompatibility of (A) with (R), to *decline to affirm* (A) is to *decline to reject* (R). He should therefore *decline to reject* (R).

This combination of methodological attitudes is additional to, yet fully compatible with, Quine's reasoning in paragraph two of section 6, as I explained it above. It also reveals a grain of truth in Putnam's later contributions to the discussion: even if we cannot *now* see how we could judge that a sentence *S* is false without changing its meaning to the point where we would no longer interpret it homophonically into our new theory, there is no guarantee that we will not *later* judge that *S* is false while still translating our previous uses of *S* into our (revised) theory homophonically.

Finally, it is also crucial to see that on the reading I sketched above, Quine is not (and should not be) committed to the claim that 'for any

view we can imagine circumstances in which we would give it up'
(Harman, 1967: 132). Quine is committed, instead, to (P), which is not a
psychological claim about what we can imagine giving up, or retracting,
but a methodological principle. Even if we are not now able to imagine
a circumstance in which we would retract a particular statement of our
current theory, we can see that no statement we now accept is guaranteed to be part of every scientific theory that we will later come to
accept. To accept this is not to make a psychological claim about what
we can now conceive, but to acknowledge both our fallibility and our
commitment to retracting any statement we now accept if and when, in
our pursuit of truth, we find it best to do so.

Acknowledgements

I presented earlier versions of this paper at the Society for the Study of
Analytic Philosophy (SSHAP) conference in Montréal in May 2014, the
Ohio State History of Analytic Philosophy conference in Dubrovnik in
June 2014, and the 'Quine and His Place in History' conference in Glasgow
in December 2014. For challenging and helpful comments at these confer-
ences, I thank Frederique Janssen-Lauret, Jeremy Heis, Peter Hylton, Mark
Kaplan, Kirk Ludwig, Andrew Lugg, Stewart Shapiro, Tom Ricketts, Sander
Verhaegh, and Alan Weir. I thank Hilary Putnam for an email in which
he explained to me that when he wrote 'The Analytic and the Synthetic',
he did not take himself to be expanding on or interpreting Quine's view
of belief revision in section 6 of 'Two Dogmas of Empiricism', but to be
offering his own related but different view of belief revision. I especially
thank Yemima Ben-Menahem, who urged me to search for more textual
evidence in support of my interpretation. I was fortunately able to find
such evidence in time to present it in this version of the paper.

Notes

1. Quine uses the word 'retract' (and also 'rescind') when discussing holism in
    Quine 1992: 14–15.
2. I say 'in effect' because Putnam does not use the notions of rational revision
    or homophonic translation in Putnam 1962. He did nevertheless implic-
    itly defend something like (R) in that paper, and in many subsequent ones.
    He explicitly talks about rational revision and homophonic translation in
    Putnam 1979.
3. Although Quine does not use the word "confirmation" in the first paragraph
    of section 6 of 'Two Dogmas', he does use the word in the second paragraph,
    where he presents himself as drawing consequences from the view sketched

in the first paragraph. He also uses the word 'confirmation' in Quine, 1960: 63–64, in the chapter in which he takes himself to be developing the account of confirmation sketched in 'Two Dogmas'.

4. There is a passage in Quine's 1936 paper 'Truth by Convention' in which he says that the statements of logic and mathematics count are 'destined' to be part of any scientific theory we later accept:

> There are statements which we choose to surrender last, if at all, in the course of revamping our sciences in the face of new discoveries; and *among these there are some which we would not surrender at all, so basic are they to our whole conceptual scheme.* Among the latter are to be counted the so-called truths of logic and mathematics... *these statements are destined to be maintained independently of our observations of the world.* (Quine, 1976: 102; emphasis added)

Is Quine claiming here that truths of logic and mathematics that we currently accept are *guaranteed* to be part of every scientific theory that we later come to accept? If so, he is thereby rejecting (P), and his argument in section 6 of 'Two Dogmas' represents a change in view. Another possibility, more likely, I believe, is that in the above passage Quine is expressing his *confidence* that we will not in fact rescind any of the statements we take to be logically true. This is a prediction, and likely a correct one, not a methodological principle that conflicts with (P).

5. In an earlier part of his reply to Quine, Carnap tries to respond to Quine's criticisms in section 4; perhaps that is why Carnap believes that in responding to Quine's reasoning in paragraph two of section 6, he can take for granted his method of clarifying 'analytic' in terms of semantical rules.

6. In the next sentence Grice and Strawson offer another formulation of the same 'point of substance' that presupposes that some statements are analytic, and they remark that their second formulation of the point is expressed 'in terms [Quine] would reject'. (Grice and Strawson, 1956: 157) This implies that in their first formulation they intend to use the word 'conceptual scheme' in a way that Quine can accept, hence, presumably, in the way he uses it in 'Two Dogmas', as, for example, in Quine, 1953a: 44.

7. Chalmers 2011 makes this aspect of the standard interpretation explicit by setting aside, for the sake of his reconstruction and evaluation of Quine's argument in paragraph 2 of section 6 of 'Two Dogmas', every other argument that Quine offers in the paper. It was only when I read Chalmers's extreme version of the standard interpretation and tried to evaluate it that I began to realize how badly it and other less extreme versions of the standard interpretation, such as the one presented by Grice and Strawson, fit the text.

8. Grice and Strawson also challenge Quine's argument that synonymy and analyticity cannot be defined in terms of confirmation, and propose an account of their own, according to which 'two statements are synonymous if and only if any experiences which, on certain assumptions about the truth values of other statements, confirm or disconfirm one of the pair, also, on the same assumptions, confirm or disconfirm the other to the same degree' (Grice and Strawson, 1956: 156). In Quine, 1960: 64–65, Quine explains why their efforts to define synonymy and analyticity in terms of confirmation fail.

9. Against this, Timothy Williamson argues that 'Someone may believe that normal human beings attain physical and psychological maturity at the age

of three, explaining away all the evidence to the contrary by *ad hoc* hypotheses or conspiracy theories' (Williamson, 2007: 85). Perhaps there are better examples, however. My points in the text do not depend on the success of Grice and Strawson's example (2).

10. He also cites Quine, 1963. See Quine, 1960: 67, note 7.

11. In his 1963 reply to Quine, Carnap writes that 'analytic sentences cannot change their truth-value. But this characteristic is not restricted to analytic sentences; it holds also for certain synthetic sentences, e.g. physical postulates and their logical consequences' (Carnap, 1963: 921). Quine's response to the argument from incomprehension therefore highlights, in a very different way – without conceding that any statements are analytic – Carnap's view that the question whether we can coherently change our judgment about the truth-value of a statement is not the key to understanding analyticity.

References

Carnap, R., & Smeaton, A. (1937 [1934]) *The Logical Syntax of Language* (London: Kegan Paul, Trench, Trubner & Co).

Carnap, R. (1963) 'W. V. Quine on Logical Truth'. In P.A. Schilpp (ed.) 1963, *The Philosophy of Rudolf Carnap*. La Salle, IL: Open Court, pp. 915–922.

Chalmers, D. (2011) 'Revisability and Conceptual Change in "Two Dogmas of Empiricism"', *Journal of Philosophy* CVIII, 8: 387–415.

Grice, H.P., & Strawson, P.F. (1956) 'In Defense of a Dogma', *Philosophical Review,* 65(2): 141–158.

Harman, G. (1967) 'Quine on Meaning and Existence, I. The Death of Meaning', *The Review of Metaphysics*, 21(1): 124–151.

Harman, G. (1994) 'Doubts about Conceptual Analysis'. In M. Michael & J. O'Leary-Hawthorne (eds.) 1994, *Philosophy and Mind: The Place of Philosophy in the Study of Mind*. Dordrecht: Kluwer, pp. 43–48. Reprinted as chapter 6 of Harman 1999. Page references in the present paper are to Harman 1999.

Harman, G. (1999) *Reasoning, Meaning, and Mind* (Oxford: Clarendon Press).

Juhl, C., & Loomis, E. (2010) *Analyticity* (New York: Routledge).

Putnam, H. (1962) 'The Analytic and the Synthetic'. In H. Feigl & B. Maxwell (eds.) 1962, *Minnesota Studies in the Philosophy of Science*, vol. III. Minneapolis: University of Minnesota Press, pp. . Reprinted as chapter 2 of Putnam 1975.

Putnam, H. (1975) *Mind, Language, and Reality*: Philosophical Papers, volume 2 (Cambridge: Cambridge University Press).

Putnam, H. (1979) 'Analyticity and Apriority: beyond Wittgenstein and Quine' In P. French et al., (eds.) 1979, *Midwest Studies in Philosophy*, volume IV. Minneapolis: University of Minnesota Press, pp. . Reprinted as chapter 7 of Putnam 1983.

Putnam, H. (1983) *Realism and Reason*: Philosophical Papers, volume 3 (Cambridge: Cambridge University Press).

Quine, W.V. (1950) *Methods of Logic,* first edition (New York: Henry Holt and Company, Inc).

Quine, W.V. (1953a) 'Two Dogmas of Empiricism', in W.V. Quine 1953, *From a Logical Point of View. Cambridge, Mass: Harvard University Press*, pp. 20–46.

Quine, W.V. (1953b) *From a Logical Point of View* (Cambridge, Mass: Harvard University Press).

Quine, W.V. (1959) *Methods of Logic*, revised edition (New York: Holt, Rinehart and Winston, Inc).

Quine, W.V. (1960) *Word and Object* (Cambridge, Mass: MIT Press).

Quine, W.V. (1963) 'Carnap and Logical Truth'. In P.A. Schilpp (ed.) 1963, *The Philosophy of Rudolf Carnap*. La Salle, IL: Open Court, pp. 385–406.

Quine, W.V. (1972) *Methods of Logic*, third edition (New York: Holt, Rinehart and Winston, Inc).

Quine, W.V. (1976) *The Ways of Paradox*, revised and enlarged edition (Cambridge, Mass: Harvard University Press).

Quine, W.V. (1982) *Methods of Logic*, fourth edition (Cambridge, Mass: Harvard University Press).

Quine, W. V. (1986) *Philosophy of Logic*, second edition (Cambridge, MA: Harvard University Press).

Quine, W.V. (1992) *Pursuit of Truth*, revised edition (Cambridge, Mass: Harvard University Press).

Williamson, T. (2007) *The Philosophy of Philosophy* (Oxford: Basil Blackwell).

Wittgenstein, L., Pears, D.F., & McGuinness B.F. (1961 [1921]) *Tractatus Logico-Philosophicus* (London: Routledge and Kegan Paul). First published German in *Annalen der Naturphilosophie*, 1921.

11
Meta-Ontology, Naturalism, and the Quine-Barcan Marcus Debate

Frederique Janssen-Lauret

1 Introduction: objects, theories, and epistemology

Quine's ontological adage, 'to be is to be the value of a variable' (Quine, 1939a: 708; Quine, 1948: 32), played a pivotal role in the revival of metaphysics which gathered momentum in the second half of the twentieth century. Yet twenty-first century critics of Quine tell us he is a narrow-minded pragmatist whose naturalism consists in restricting metaphysics to the trivial and uninteresting pursuit of existence questions, especially existence questions implied by the physical sciences. Investigating Quine's views in their proper context reveals that this is a misinterpretation; his work on ontology is meant to give us a systematic way of identifying where a theory provides good reason to believe in something. He makes explicit that the influence of his pragmatism and naturalism on his meta-ontology is not to shrink the role of metaphysics and truth in our theories, or to restrict the content of available theories to the physical. It is to give a theory-transcendent criterion of where theories, scientific and philosophical, assume the existence of something: where it is posited as the occupant of some indispensable theoretical role. Modern critics try to locate Quine's naturalism where it does not belong, and as a result dismiss his view as far less interesting than it really is, as I will argue in Sections 2 and 3.

Still, in Sections 4–6 we see that Quine's particular form of naturalism does, after all, place some unhelpful limits on his meta-ontology. The way in which it constrains the available options is entirely different from the way twenty-first century metaphysicians imagine it to be; he is in no danger of collapsing into anti-realism, deflationism, or triviality. An unfairly neglected debate shows that at least one philosophically interesting metaphysical position is, nevertheless, ruled out by

Quinean meta-ontology. Ruth Barcan Marcus, whom Quine thought of as a worthy adversary, at first glance appears to be much closer to Quine than modern Carnapians or neo-Aristotelians – like him she is a realist and a naturalist, and she is also a committed nominalist. Barcan Marcus' debate with Quine has not been given the attention it deserves because their meta-ontological discussions occur along the way in the course of their better-known dispute on modality. Her decisive victory over Quine on the issue of necessary identity is now widely appreciated, and she is also beginning to be recognized as the originator of the direct reference theory of names within analytic philosophy (Barcan Marcus, 1961). But what has been missed until now is that Barcan Marcus also presents, just like Quine, a criterion of ontological commitment, complete with a characteristic canonical language of regimentation: a fully-fledged meta-ontology. Unlike Quine, she takes proper names, and not variables, to indicate ontological commitments. Close reading of the relevant texts reveals some mutual misunderstandings, leading to Barcan Marcus' views, though just as interesting as Quine's, being misinterpreted and ultimately forgotten. I now intend to revive her position as a viable meta-ontological alternative, especially attractive to contemporary adherents of her direct reference theory. Tracing the steps of their debate and reciprocal misreadings makes clear that the root of their differences is epistemology. Barcan Marcus believes in knowledge by acquaintance, codified in her canonical language by proper names. Quine does not, and accordingly bans names from his.

2 Quine's meta-ontology and its relationship to his naturalism

2.1 Quine's naturalism, his holism, and the continuity between science and philosophy

The centrality of Quine's naturalistic outlook to his meta-ontology is often misunderstood. His meta-ontological ambitions are of a broad and philosophical character: his criterion of ontological commitment provides a strategy for answering weighty questions in the intersection of logic, metaphysics, and epistemology. What kinds of entities do I have good reason to believe in, given my best theory? How to make the ontology of a theory maximally clear based on its logical form? What objects do alternative theories give others licence to believe in? Although Quine was critical of several aspects of traditional metaphysics, the historical record makes clear that his intention was to preserve what was worth saving about it. Like Quine's own ontology, what has come to be called his *meta-ontology* also has an explicitly naturalistic tenor: he intended to safeguard metaphysics from positivistic attacks by showing it to be continuous with natural science.

Analytic philosophers' interest in ontology faltered with the logical positivists, who thought of existence questions as uninteresting, meaningless, or misguided – trivial analytic consequences of a chosen language form at best, at worst ill-formed and not answerable at all (Carnap, 1950: 25). Quine cast doubt upon the underlying assumption that language is separable into a trivial, merely conventional part and a factual part. According to him, all statements of a theory are true or false because of a combination of fact and convention; no truths are trivially true because of language alone. Confirmation is not a matter of holding up individual theoretical statements against reality and assessing whether they match reality, if they are empirical, or whether they are just not in the business of trying to match reality. Rather, a theory, a set of sentences closed under consequence, touches upon reality as a whole, and is confirmed or disconfirmed along with everything that follows from it, including existential quantifications (Quine, 1951a: 69–72). As there is no clear boundary between mere useful convention and statements that are true because of the way the world is, the dividing line between a theory's empirical claims and its logical, mathematical, or philosophical conventions dissolves: 'I see metaphysics, good and bad, as a continuation of science, good and bad' (Quine, 1988: 117). Existentially quantified sentences are confirmed along with others that are more closely linked to observations, because science and philosophy share a concern with what we have good reason to believe about the world. And one vital component of this concern is what kind of objects we have reason to believe the world contains based on a given theory.

2.2   Quine on theory-building and ontology

Quinean ontological commitments always take the logical form $\ulcorner \exists x \varphi(x) \urcorner$: they contain a variable, some description containing some of the predicates of the theory that is supposed to hold of the value of that variable, and an existential quantifier to bind the variable. But why? In the late thirties to early fifties, Quine recommends 'translating' disputed existence claims into quantificational form to clarify them and remove any ambiguity. 'Monotheists and atheists now need disagree only on the truth values of statements such as $[(\exists y)(x)(x = y. \equiv god\ x)']$, not on questions of meaningfulness' (Quine, 1940: 150; see also 1939: 705–708). But the notion of 'translation' used here stands in need of clarification, too, as Alston (1958: 12–13) and Cartwright (1954: 3), for instance, point out;

from the late fifties onwards, perhaps under pressure from these quarters, Quine begins to develop an explanation that relies more on his philosophy of language and holist epistemology (Quine, 1957–1958). Quine believes that acquiring a theory, whether in infancy, or via translation, or in a scientific setting, is just to acquire a language, grown out of observations. The very early stages of theorising are composed of nothing except observation sentences, used by the budding theorist to label features in her experience: 'Tree'; 'Green'; 'Rose'; 'Red'; 'Rabbit'; 'Furry', *et cetera* (Quine, 1960: 92). From individual observations nothing else follows. All we can do with observation sentences at this stage is venture them and see what reactions we encounter. The first additions to the emerging theory are 'yes' and 'no', based on the reactions of assent and dissent. The nascent theorist is then in a position to learn to use truth-functional operators such as negation, conjunction, disjunction, and the conditional (Quine, 1960: 57–59). Thus far all of the theory is empirically conditioned, directly based on experience. It is possible for theories never to develop beyond this point. Such theories remain, structurally speaking, on the level of sentential logic, and have no ontology (Quine, 1979). There are only atomic sentences and truth-functional connectives to link them.

Still, further developments have great explanatory benefits. We are able to distinguish much more fine-grained kinds of evidence if we are able to discuss and explain which observations frequently coincide and why. Our vehicle for locating patterns within the observations is the pronoun, the introduction of which turns observations into predicates, and leads the language to be enriched with further logical vocabulary: quantifiers binding the variables. Adding these resources amounts to a language with the expressive strength of first-order logic. The language learner will begin to use pronominal expressions when she notices significant intersections in her observations. The difference between intersections and mere conjunctions is marked by inserting a pronoun where observations coincide in an interesting way. 'Green. And Tree' is true in the presence of a green field and a copper beech. 'This is green and it is a tree' says something more: that green and tree persistently overlap here. The ontological vocabulary pinpoints these intersections. Reification begins when, in response to recurrent evidence of such intersections, we posit an object as a likely explanation of the pattern of overlap (Quine, 1992: 24). Pronouns signal an increase in explanatory capacity of a growing theory. The introduction of pronouns into the observation sentences imposes a structure on the sentences that was not there before. Previously an undivided whole, they now have a pronominal part and a predicative part. The predicative part is what is left over from the observation sentence. Observation sentences are feature-placing (cf. Strawson, 1959: 212), not attributive: they do not contain a word for an object and a word that attributes something to the object. The addition of variables divides the language into expressions that purport to say *what there is* and those that purport to say *what is true of* those beings. The former are the ontological expressions, and the latter the ideological expressions. The ideology comprises all the characteristics ascribed to the beings: what can be truly said of them and how they are related.

2.3   Naturalism, pragmatism, formalisation, and realism

Our best theory, couched in the clearest, least ambiguous language we can muster, will tell us exactly what it says there is: the values of its pronominal expressions. Quine is frequently criticised, though sometimes also praised (van Inwagen, 1998: 235–237), for maintaining that the existential quantifier is closely related to ordinary-language 'there is'. Although Quine does occasionally say that existential quantification is just a tidied-up version of ordinary-language 'there is' or 'exists' (Quine, 1940: 65–71; 1969: 94), at times he also expresses a much more pessimistic attitude towards the idea of extracting an ontology from ordinary-language utterances: 'Ontological concern is not a correction of a lay thought and practice; it is foreign to the lay culture, though an outgrowth of it' (Quine, 1981b: 9). But according to Quine all posits, naive or sophisticated, stem from the introduction of an entity on the intersection of continuous observations by means of a pronoun; if a theory lacks this quantificational form, it should be imposed upon it to work out its ontology. The theory should be regimented (Quine, 1960, ch. 5): translated – radically, following the stages outlined above – into the canonical language of first-order logic. The process of regimentation also allows us some strategies for translating out unwanted ontological commitments. Suppose we would like to deny that there are φs, but we find ourselves making an utterance which apparently asserts or implies the existence of φs. There are three options for regimenting our theory in such a way that it avoids commitment to φs. First of all, we could, in Quine's words, 'take an attitude of frivolity' (Quine, 1953: 103), or deny that the utterances φ-language occurs in are properly part of a theory at all. Second, we could eliminate: remove φ-language from our theory altogether. The third option is reduction by paraphrase, or providing a template for translating all occurrences of φ-language into statements that make no commitment to φs, in the process producing an improved, equally explanatory but more parsimonious new theory.

As the theory progresses after its first attempts at reification, more sophisticated posits and methods are likely to ensue. Scientific standards of rigour will be formulated within the theory. It will be held up to and judged by standards of theory choice, such as simplicity, testability, and fecundity (Quine, 1955 [1966]: 247). Its posits, too, will be re-evaluated. Questions will arise about which intersections of observations are the significant ones, the ones that call out for a posit. These questions are not answerable on purely empirical grounds (Quine, 1960: 51–55). As soon as we begin to reify, we introduce indeterminacy of translation, too. There is no way of pointing directly to an object. Instead they are marked out on the foci of interlocking observations, but not on *all* foci of interlocking observations – just the ones where we take ourselves to have the best reason to suppose an object is present as a plausible explanation of the way they intersect. How to divide the phenomena into theoretical roles that call for entities is a matter of weighing up the different options and judging them by our current best standards of theory choice: what is the simplest, most explanatory theory that best fits the data? Pragmatism, Quine-style, means that we have to keep updating our theory, as well as retain a willingness to readjust the standards by which we judge theories, though there is no God's-eye perspective from which we can assess our current theory. Any theory is such that we have to keep improving it from within, including the re-evaluation of which observations collectively amount to an explanatory role that demands the presence of an entity.

Quine has given us a template to work out, for any theory which purports to say true things about the world out there, what entities it takes that world to contain. Once all its theoretical contexts have been clarified, collected together, and closed under consequence by the process of regimentation, the existence claims will reveal what objects there are, according to this theory, and what explanatory roles they play. So why is his meta-ontology so frequently derided by twenty-first century philosophers as uninteresting, trivial, or covertly anti-realist?

3 Quine the metaphysician: rebutting some misinterpretations of Quine as anti-metaphysical

3.1 Quine the realist holist

Quine's willingness to revise any aspect of a theory in the light of new evidence, and his unwillingness to admit a higher authority than our current best theory, should not be read as willingness to relinquish realism about truth, reference, or ontology. Recently some critics have portrayed Quine as an anti-metaphysical pragmatist (Schaffer, 2009: 349; Price, 2009: 326). But this reading is hard to square with Quine's own work. The pragmatism he commits himself to is, rather, epistemological, devoted to working within our best theory and improving it from the inside, and embracing porous boundaries between subject areas. Quine often expresses some measure of skepticism about old-fashioned metaphysics, but one of his meta-ontological aims is repackaging ancient debates into exciting new logical forms, the better to debate them anew. He points out that his efforts are continuous with those of earlier generations of philosophers, salvaging the kernel of truth contained within the old metaphysics and discarding what's been superseded: 'Though no champion of traditional metaphysics, I suspect that the sense in which I use this crusty old word ["ontology"] has been nuclear to its usage all along' (Quine, 1951a: 66).

Price portrays Quine as committed to anti-metaphysical pragmatism and unable to make a clear distinction between truth and usefulness. In defence of this interpretation he cites only the final sentence of 'On Carnap's Views on Ontology': 'Carnap maintains that ontological questions...are questions not of fact but of choosing a convenient scheme or framework for science; and with this I agree only if the same be conceded for every scientific hypothesis' (Quine, 1951a: 72). In context, that reading of the sentence becomes hard to sustain. The final sentence is immediately preceded by a reference to Quine's rejection of the analytic-synthetic distinction (Quine, 1951b), countering Carnap's division of existence questions into internal (factual or analytic) and external (merely conventional) with his own argument that there are neither purely factual statements nor purely analytic or conventional statements wholly devoid of factual content (Quine, 1951a: 72; see also MacBride and Janssen-Lauret, forthcoming 2015, section 3). Even Carnap's own paper supports my reading: 'Quine does not acknowledge the [internal/external] distinction I emphasise above, because according to his general conception there are no sharp boundary lines between logical and factual truth, between questions of meaning and questions of fact' (Carnap, 1950, footnote 5).

Quine can frequently be found expressing the view that a useful theory is one we have good reason to hold true (a view which Price decries as 'surely a misinterpretation' of Quine (Price, 2009, footnote 4)):

> 'Everything to which we concede existence is a posit from the standpoint of a description of the theory-building process, and simultaneously real from the standpoint of the theory that is being built. Nor

let us look down on the standpoint of the theory as make-believe, for we can never do better than occupy the standpoint of some theory or other, the best we can muster at the time' (Quine, 1960: 22; also see Quine 1981a and Ben-Menahem's discussion of Quine 1981a in her paper in this volume).

Price might object that it is a further question whether this gives us good reason to believe that the values of its variables exist, since he reads Quine as a minimalist about truth (Price, 2009: 325). But Quine is not a minimalist about truth. He endorses disquotationalism, the Tarskian doctrine that a materially adequate truth-definition for a language L which does not contain its own truth predicate must entail all T-biconditionals for that language, formulated in a stronger metalanguage which contains L's semantic vocabulary (Tarski, 1956 [1933]: 188). Minimalism about truth is disquotationalism plus one additional thesis: that there is nothing more to be said about truth than is expressed by the T-biconditionals (Horwich, 1990). Truth, for the minimalist, need not involve contact with objects. Quine rejects that thesis (as did Tarski before him), stating that an atomic sentence is true iff the value of its variable – that is, an object it stands for, an element of some domain – satisfies its predicate: 'Tarski's satisfaction relation has to do with objective reference, relating open sentences as it does to sequences of objects that are values of the variables' (Quine, 1976 [1970]: 318).

3.2   Quine against triviality

A neighboring misreading of Quine portrays him as an anti-metaphysical pragmatist whose meta-ontology restricts the role of metaphysics to existence questions, which are philosophically uninteresting because they follow trivially from a theory (Schaffer, 2009: 358). Quine himself must take some blame for this interpretation, as he asserts in the late forties that ontology is trivial to the conceptual scheme (Quine, 1948: 29). But this vestige of Carnapianism was soon abandoned, since it is incompatible with Quine's semantic holism. Any part of a language is potentially revisable under sufficient theoretical pressure, including its definitions, mathematics, and logical laws (Quine, 1951b: 40). So nothing follows trivially from anything.

Schaffer cites Hofweber (2005) puzzling over the contrast between the deep nature of metaphysical questions and the alleged triviality of existence questions, concluding that the culprit is Quinean meta-ontology (2009: 361). To say this is to overlook both Quine's opposition to taking existence questions as trivial in the first place and the Quinean inseparability of ontology and ideology.[1] A posit, for Quine, is always introduced because it fills some explanatory role phrased in terms of the theory's ideology. Every existence question which follows from some attempt at a best overall theory of the world is in fact an interesting, non-trivial consequence of it, because there is no statement of any theory so sacrosanct that it cannot be reinterpreted in a way that yields a different set of existential consequences.

3.3   Quine's naturalism, his own ontology, and his meta-ontology

A related complaint is that Quine fails to see that while metaphysical existence questions are trivial, all non-trivial, interesting existence questions – e.g. the Higgs boson – 'of course [have] nothing to do with metaphysics' (Tahko, 2011: 28). Contemporary metaphysicians pushing this complaint often want to return to the *a priori* categories of traditional Aristotelian metaphysics. Neo-Aristotelians contend that we know what an object is, and how objects in general must fit together, in advance of knowing any individual object, empirical investigation being restricted to which individual objects there are, and how they are (Lowe, 2006: 5). Quine certainly opposes such a sharp separation between *a priori* metaphysical questions and *a posteriori* empirical questions – but not, as the neo-Aristotelians think, because his naturalism leads him to formalize physics and resign metaphysics to the task of working out its existential consequences (Schaffer, 2009: 366–367). This is to mistake Quine's own ontology for his meta-ontology. Quine's own preference for physicalism does not restrict the range of theories to which his ontological method can be applied; in fact, he is explicit from his earliest writings on meta-ontology that its interest lies in regimenting and clarifying alternative views, including more traditional philosophical ones. He mentions using quantificational form to elucidate the views of realists about propositions (Quine, 1939: 708), theists (Quine, 1940: 150), and mathematical Platonists (Quine, 1944: 161). And the fact that he carries on attempting to regiment and assess others' existential assumptions after his introduction of the doctrine of ontological relativity (Quine, 1968) – for instance, his worries about the criteria of identity of possible worlds (Quine, 1976) – indicates that this doctrine has not, like Schaffer says (2009: 349, 372), turned him into a deflationist about ontology. None of the contemporary arguments against Quine get him quite right; none of them indicate that he – or his contemporary followers – should give up his meta-ontology.

My interpretation of Quine reveals a much more interesting figure whose criterion of ontological commitment remains relevant, one whose naturalism inspired him to attempt to find a rigorous criterion, applicable to any theory, for having good reason to believe in something based on that theory. Quine the naturalist sought such a criterion in the practices of science, where posits are justified by their fulfilling some designated explanatory role. Quine the pragmatist applied this insight to philosophy, since he viewed the boundary between science and philosophy as porous in the first place. With this in place we will now see that, although contemporary critics are looking for flaws in all the wrong places, there is, nevertheless, a flaw in Quine's meta-ontology. He thinks that his criterion can account for *all* kinds of existential assumptions, but there is one significant type that he overlooks: ontological commitment via direct reference. This is because Quine's global holism is so reliant on descriptions that direct reference is difficult to make sense of.

4   Epistemological and metaphysical corollaries of Quine's meta-ontology: names and identity

4.1   All knowledge is general knowledge

According to Quine, all theories introduce objects into their structure in the same way, and for the same purpose: they are tentatively put forward as a best explanation of a pattern of intersecting observations. Every object which belongs to the ontology of some theory, according to him, is there for a distinct explanatory reason: the theory has a need for an entity which satisfies some description φ. Our only access to objects is descriptive, by considering the object's role in our best theory. His holism implies that all of our knowledge is in principle general knowledge. So Quine's view implies a further holism-inspired thesis. We know objects only *qua* solutions to puzzles on how to explain the phenomena. Ontological commitments are incurred only to objects insofar as they fall under some open formula φ of the theory: objects-*qua*-φ. Since we keep moving step by step towards better theories, we gradually replace our posits with better posits, too: atoms with sub-atomic particles, space with space-time, *et cetera*. Quine's meta-ontology allows us to make sense of these developments. Any object-*qua*-φ is potentially dispensable. When we readjust our explanations in response to empirical developments, we drop the old posits and posit objects that answer to our new, improved descriptions.

4.2   Criteria of identity

These descriptive explanatory roles impose criteria of identity on objects. Quine's meta-ontology implies that all objects are subject to indiscernibility according to the predicates of the theory. Beyond the use of predicates, there is no theoretical apparatus available to distinguish them from each other. Hypostatizing a posit on the intersection of observations in the first place is a significant theoretical imposition for Quine, let alone a single self-identical thing being identified in different contexts. Any particular posit may be dispensable. To decide whether the existence of something is really implied by the theory, the entire wealth of theoretical resources it provides must be invoked. Criteria of identity gleaned from a complete theory are a key component of such decisions: 'Our venerable theory of the persistence and recurrence of bodies is characteristic of the use of reification in integrating our system of the world. If I were to try to decide whether the penny now in my pocket is the one that was there last week, or just another one like it, I would have to explore quite varied aspects of my overall scheme of things, so as to reconstruct the simplest, most plausible account of my interim movements, costumes, and expenditures' (Quine, 1992: 24).

If an object can only be admitted into the ontology by playing some indispensable explanatory role couched in terms of the predicates of the theory, there will never be any reason to admit two objects which satisfy all and only the same predicates of that theory. There is no distinct explanatory role for the second object that the first doesn't already discharge. This implies that there is no role for identity beyond indiscernibility-according-to-the-theory. Quine proposes that ' "φxy" meets all the formal requirements of "$x=y$"' iff $(\forall x)\varphi xx$ and $(\forall x)(\forall y)(\varphi xy$ & $(...x... \rightarrow ...y...))$ (Quine, 1976 [1961]: 180, notation modernized.) He admits that his 'serviceable facsimile' for identity does not amount to the usual identity relation, the relation which partitions the domain into singleton equivalence classes (Quine, 1970: 64), or simply *being the same thing* (Quine, 1960: 114–118). There may be differences between objects that the predicates of the language fail to capture. But he avers that from within the language, such differences are inexpressible; 'identification of indiscernibles' (Quine, 1953 [1950]: 71; 1947a: 75) is the best we can do. Nor can we appeal to a stronger language to distinguish the two. Quine's pragmatist naturalism implies that we must always work within our best theory.

Where two posits satisfy all and only the same open formulae, they count as the same object for the purposes of that theory. We only have good reason to believe in one object per explanatory role φ – after all, φ is specific enough to include exact spatiotemporal coordinates and the like – so objects-*qua*-all-and-only-the-same-φs are, according to Quine, to be identified. Quinean regimented theories cannot have two distinct but indiscernible values of variables in their domains. To make the cut for being in the domain, an object has to contribute a distinct explanatory role, which means that it satisfies some open formula no other object in that domain satisfies.

4.3 Dispensing with proper names

It appears counterintuitive that a language can speak of two distinct objects without being able to tell them apart in any way. Could we not refer to each of them by name to distinguish them? Not according to Quine, who proposes to translate away all occurrences of proper names in the process of regimentation. In the 1930s, Quine had thought of names as ontologically committing: 'To ask whether there is such an entity as roundness is...asking whether this word is a name or a syncategorematic expression' (Quine, 1976 [1939]: 197);[2] '"Pegasus" is not a name in the semantic sense, i.e.,...it has no designatum' (Quine, 1939: 703).[3] He explicitly abjures their use in ontology in the forties, saying they are dispensable in favor of bound variables (Quine, 1948: 32). Any name *a*, he claims, can easily be transformed into a predicate *Ax* which applies only to the object formerly named *a* if it applies to anything (Quine, 1970: 25–26). 'Europe' becomes 'the unique object which europizes' (Quine, 1940: 149). Quine returns to this argument several times over the years, but is frequently less than clear about whether '*x* europizes' is equivalent to '*x* = Europe' (as he says in Quine, 1970: 25–26), or to 'repars[ing]' the singular term into a general term (Quine, 1976c [1954]: 238). The former, as Barcan Marcus objects, does not eliminate the name so much as recycle it (Barcan Marcus, 1993b [1985]: 211). The latter puts a proper name in a place where only a predicate can go – the sort of move Quine fervently deplores when it is perpetrated in reverse. Anyone who would dare put predicates (Quine, 1960: 119–120) or propositions (Quine, 1943: 120) in name-position is berated for butchering logical syntax. But whether 'europizing' is acceptable or not matters very little in the end. Quine's real goal is to dispense with names in favor of a 'pat translation into a descriptive phrase...along familiar lines' (Quine, 1948: 27), that is, using the ideology of the theory. His view collapses into straightforward descriptivism about names, because the requirement that every entity must be subject to a criterion of identity applies to naming too. Any language that meets his standards would have assigned names by means of a prior description in the first place.

To assign a name to an object we must first have singled it out by means of a descriptive phrase. Otherwise, we could never be in a position to identify the object we just named with some object encountered in a different theoretical context. Suppose we name a river by ostension; the bearer of that name will be subject to the criterion of identity for rivers, not those for collections of water particles or spatiotemporal zones. So the ostended object must first be described as a river. '[Ostention] leaves no ambiguity as to the object of reference if the word 'river' itself is already intelligible' (Quine, 1953 [1950]: 67). Later, when wondering whether some observed flowing body of water deserves the same appellation as the subject of our ostension, we must first make clear under what circumstances two observed entities count as *the same river,* i.e. are subject to the criterion of identity for rivers, or what Quine calls 'river kinship': 'the introduction of rivers as single entities, namely, processes or time-consuming objects, consists substantially in reading identity in place of river kinship. ... The imputation of identity is essential, here, to fixing the reference of the ostension' (Quine, 1953 [1950]: 66). Quinean theories only have ontological commitments to things their ideology can describe – proper names themselves cannot introduce an ontological commitment to an entity. A theory cannot be committed to an individual *qua* individual, independently of how it is described. But this is precisely the kind of ontological commitment that Quine's interlocutor Ruth Barcan Marcus advocates.

5   Barcan Marcus' name-based meta-ontology

5.1   Barcan Marcus' name-based criterion of ontological commitment

Like Quine, Barcan Marcus is naturalistically inclined, professing a distaste for abstract objects. But their meta-ontologies are diametrically opposed, because of a difference in epistemology (Barcan Marcus, 1978: 358–359). Although Barcan Marcus shares Quine's affinity for physical objects and distaste for propositions, properties, and other abstracta[4] like numbers and sets, and an unwillingness to ascribe referential status to predicates and logical operators, her nominalism has a very different epistemological motivation: she is a foundationalist. She takes as a point of departure the idea that when we interact with the world, we encounter individuals, which we come to know by acquaintance. Our minds reach out to concrete objects like other people, organisms, artefacts, and lifeless matter, and our command of names allows us to stick a label on any such concrete object: a tag (Barcan Marcus, 1961: 310). Tags' semantic function is to stand for things, while other forms of language – predicates, logical operators, mathematical discourse, the language of moments in time or events, talk of propositions, and fiction (Barcan Marcus, 1972) – have no subject matter for them to refer to. So they are mere language, a *flatus vocis*. She expresses puzzlement at Quine's decision to let variables 'bear the burden of reference', preferring 'alternative analyses for locating references in an interpreted language...names and their relation to nameable objects' (Barcan Marcus, 1978: 359).

Barcan Marcus can be read as proposing a fully-fledged criterion of ontological commitment with its own attendant canonical language of regimentation to complement it, summed up as a slogan to rival Quine's: 'to be is to be the referent of a tag'. For Barcan Marcus, it is not variables, but directly referential proper names, that betoken ontological commitment. Not every apparent proper name in English will go into her regimented language as a tag. Her epistemology implies that concrete individuals can be assigned tags, but we have no good reason to believe in abstracta and therefore our predicates, logical constants, numerical terms, propositional locutions, abstract nouns, and fictional empty names are syncategorematic, and will be rendered as non-referring expressions in her canonical language.

Existence claims, and the variables contained within them, are not ontologically committing for Barcan Marcus. She combines her tag theory of reference with a substitutional interpretation of the quantifiers. Where Quine attempted to reduce apparently ontologically committing uses of proper names to variables, Barcan Marcus' solution is an exact mirror image of Quine's: quantifiers are reduced to names. Quantificational truth is explained in terms of substitution instances. Variables are relieved of the burden of reference; they are not by themselves ontologically committing, because they do not have values. They are placeholders for substituends. She believes substitutional quantification to be a better match for her nominalism and a better rendition of many of our ordinary-language statements: 'There are, even in ordinary use, quantifier phrases that seem to be ontologically more neutral, as in: "It is sometimes the case that species and kinds are, in the course of evolution, extinguished". It does not seem to me that the presence there of a quantifier *forces* an ontology of kinds or species. If the case is to be made for reference of kind terms, it would have to be made, as for proper names, independently. Translation into a substitutional language does not force the ontology. Such usage remains, literally and until the case for reference can be made, *à façon de parler'* (Barcan Marcus, 1978: 359).

5.1   Barcan Marcus' canonical language of regimentation

It is fairly easy to see what Barcan Marcus' canonical language should look like; one proposal of such a language was sketched by Dunn and Belnap (Dunn and Belnap, 1968). Its lexicon would at minimum contain a finite or denumerable set of constants, a finite or denumerable set of predicates, a denumerable set of variables, the usual truth-functional operators (for instance, '¬' and '&'), the quantifiers ('∀', '∃'), and an identity predicate. There would also be a category of singular terms that are not constants, never deployed as tags. Barcan Marcus further allows for lexical items like modal operators, second order variables, or set-theoretic vocabulary, none of which are ontologically committing (Barcan Marcus, 1972). The standard syntax applies, but it must stipulate that only constants or variables can flank the identity sign, because only individuals are self-identical. The non-tag singular terms can be concatenated with other predicates, but not identity. The interpretation maps the constants to the individuals of a domain, where every constant is a tag: each is assigned to an element and each element has a name. That function may be one-one, but need not be (Barcan Marcus, 1961: 309–311).

Atomic sentences consisting of a predicate plus a tag are true when the bearer of the tag satisfies the predicate, false otherwise. All atomic sentences of the form '$a = a$' are true. Those of the form '$a = b$' are true iff the bearer of 'a' is the same individual as the bearer of 'b', false otherwise. All other atomic sentences, including those with non-tag singular terms, are assigned truth values by the interpretation. All connectives as well as the substitutional quantifiers can be assigned true or false in terms of truth alone: '$\neg p$' is true iff 'p' is false, '$p \& q$' is true iff 'p' is true and 'q' is true, and $\forall x F x$' is true iff 'Ft' is true for all terms 't'; '$\exists x F x$' is true iff 'Ft' is true for at least one term 't' (Dunn and Belnap, 1968). All substitution instances which contain non-tag singular terms as substituends – for Barcan Marcus, this would include fictional terms, higher-order terms, and terms for mere possibilia, for example – will have been assigned truth values by the interpretation quite independently of any ontological considerations, and have no bearing on the ontology. What, then, is the theory's ontology? All and only the things it makes direct reference to: all the referents of its tags. The quantifiers are ontologically inert. Barcan Marcus' conception of identity as old-fashioned sameness of thing plus her idea that direct reference commits the speaker to the existence of an object allows us to introduce ontological commitments incurred via acquaintance rather than descriptively.

6 Barcan Marcus wins on names, Quine on quantifiers

Quine considered Barcan Marcus a formidable opponent he enjoyed crossing swords with[5] and regularly credits her with significant changes of heart. His favorable review (Quine, 1947b) of Barcan's quantified modal logic (Barcan, 1947) spelled the end of his previous position that modal logic was invariably a use-mention confusion got out of hand, although he carried on believing for a few more years that modal operators are to be analyzed away in terms of analyticity (Quine, 1947c: 45; Quine, 1951b: 22). Undeterred, Barcan Marcus mounted a strong campaign against Quine's anti-modal arguments, demonstrating how his 'number of planets' argument (Quine, 1953) is invalid in her QS4 (Barcan Marcus, 1961: 313–314) and how his 'mathematical cyclist' argument[6] (Quine, 1960: 199) does not entail the conclusion he thinks it does, that modal logics must divide attributes into the essential and the accidental (Barcan Marcus, 1967). Quine kept conceding ground to Barcan Marcus on modal logic (a catalogue of her victories is detailed in Barcan Marcus, 1990). On the issue of substitutional quantification, too, she gradually convinced him that her interpretation was a respectable alternative to his. After discussions with Barcan Marcus he began to understand that her interpretation is not incoherent, but a result of dissociating the quantifier from ontological commitment. He subsequently describes it as an intelligible, though 'deviant', reading of quantification (Quine, 1970: 89–90), even providing an alternative Tarskian truth definition for it (Quine, 1976 [1970]). But Quine does not seem to have fully appreciated the epistemological dimension of Barcan Marcus' views. He never takes her endorsement of direct reference to heart, claiming the distinction between names as tags and descriptions is 'a red herring' (Barcan Marcus, 1993a: 34).

This misunderstanding, plus his failure to realize that Barcan Marcus has a criterion of ontological commitment and canonical language to rival his own, means that Quine never directly addressed her arguments for the ontologically committing status of names. Barcan Marcus' direct reference theory of proper names (Barcan Marcus, 1961), brought into the philosophical mainstream by Kripke (Kripke, 1980), now has numerous adherents. Staunch direct reference theorists might prefer to give up Quinean meta-ontology if it is as inextricably linked to his descriptivism as would appear – so ought they to embrace Barcanian ontological commitment instead?

Quine does offer some explicit arguments against having proper names as part of the canonical language, although they draw mostly upon the philosophy of logic rather than epistemology. None of them are especially effective against Barcan Marcus. Quine's first concern is to avoid the Meinongian inference that using a fictional name, like 'Pegasus', implies an existence claim by existential generalization (Quine, 1948: 26). But in Barcan Marcus' canonical language, what she calls 'E-generalisation' does not entail that something exists (Barcan Marcus, 1978: 358); only the use of a tag does. Since 'Pegasus' is not a tag, Quine's problem does not arise. Secondly, he contends that the use of a name need not incur an ontological commitment, because it can be coherently denied that the name-like phrase is a name if it fails to denote (Quine, 1951a: 67). But this does not refute Barcan Marcus, who does not believe that any use of an apparent name *trivially* implies a commitment to a bearer – after all, it is coherent to translate an apparently name-like expression into the canonical language as a non-tag term or a description, if it fails to fulfil the characteristic name-like function of singling out an object directly. The problem is not with names, but with the assumption that anything follows trivially from anything, which, we saw above, anyway relies on a dubious analytic-synthetic dichotomy.

Lastly, Quine objects that omitting names simplifies logical syntax, allowing for a single category of singular terms. A language with both constants and variables must have two such categories, since they have different syntactic properties. A variable can be appended to a quantifier symbol in order to bind all of its subsequent occurrences in a subformula; a name cannot (Quine, 1970: 26). This is true enough, but not decisive. Quine himself advocates trading simplicity of syntax for greater expressive power when he justifies embracing a first-order language over one with the structure of sentential logic. A bifurcated category of singular terms which bestows other theoretical advantages upon a language constitutes a good reason to opt for a more complex syntax.

And there are some theoretical advantages to ontologically committing proper names, principally the option to distinguish names from definite descriptions. On this point, too, Barcan Marcus and Quine continued to talk past each other over the course of several interactions. Barcan Marcus misunderstood Quine just as much as he did her, apparently taking him to be merely confused: 'It is a curious fact that Quine, who leaned on the theory of descriptions in "On What

There Is" as a solution to puzzles about nonreferring singular terms, failed to see its effectiveness in dispelling his apparent puzzles about substitutivity in modal contexts' (Barcan Marcus, 1993b [1985]: 192). This curious fact is accounted for once we see that on Quine's view, names can never rise above the level of descriptions, nor identity above the level of indiscernibility. Recall that Quinean commitments are aways made to objects-*qua*-φ, objects insofar as they satisfy some open formula of the theory. So ontological commitment to an object *a* must be regimented instead as commitment to the φ. As a result Quine cannot admit any theoretically salient distinction between '*a* = *b*' and 'the φ satisfies exactly the same open formulae as the ψ'. Barcan Marcus would counter that '*a*' and 'the φ' are very different in both meaning and semantic function: '*a* = *b*' is, if true, a necessary truth, even a logical truth, because it is arrived at by substituting co-referential terms in the logical truth '*a* = *a*' (Barcan Marcus, 1961: 308). She would also object that satisfying exactly the same open formulae is not a sufficient condition for identity. Descriptions cannot be 'strongly equated' with each other, the way directly referential expressions such as names and variables can, but only weakly (Barcan Marcus, 1961: 310–311). Barcan Marcus appears to have overlooked that Quine-style ontological commitment implies descriptivism and a weak imitation of identity. She may initially have been misled by the fact that he repeatedly tries to shelve, in his reply to her and subsequent discussion, the question of direct reference as irrelevant to modality (1976e [1961]: 181–182), into thinking that he accepted direct reference but failed to see its pertinence to the issue. Quine's use of 'socratising'-style predicates may also have led her down this path. She reads them as having the deep logical structure of '*x* = Socrates', with identity in her own, strong sense, and 'Socrates' occurring as a genuine proper name. She opposes this move on the grounds that 'being identical to Socrates is...not a general property [because such] properties...have components that refer to individuals directly' (Barcan Marcus, 1993a: 231, footnote 49).[7] These kinds of predicates covertly rely on the semantic force of direct reference in order to be meaningful at all. Therefore '[s] uch devices do not *eliminate* the name, they recycle it' (Barcan Marcus, 1993b [1985]: 211–212).

A name-based criterion of ontological commitment is more congenial to modern direct reference theorists, of which there are many, than Quine's attitude towards names. Still, Quine has the advantage over Barcan Marcus in one respect. Because her substitutional quantifier always requires terms as substituends for its truth conditions, there can be at most as many things quantified over as there are names. Names, being discrete, listable items, must be denumerable. Compared to the objectual quantifier, Barcan Marcus' quantifier is simultaneously too strong and too weak. Too strong, because if the substitutional quantifier is such that '$\forall x Fx$' is equivalent to 'Fa_0 & Fa_1 & $Fa_2 \ldots$', with a denumerable infinity of names 'a_0', 'a_1', 'a_2' ... that 'Fx' can be concatenated with, the quantifier amounts to an ω-rule. This would compel the substitutionalist to reject compactness and thereby the traditional classical consequence relation (Dunn and Belnap, 1968: 180; Shapiro, 2000, § 9.1.4.). Too weak, because since the number of names cannot be more than denumerably infinite, it cannot quantify over non-denumerable domains. But there are plausible ontologies with non-denumerable cardinalities: for instance, those that quantify over the natural numbers, or non-quantized spacetime points. Barcan Marcus' nominalist sympathies, rooted as they are in a strongly foundationalist epistemology of the encounterability of individuals through acquaintance, lead her to profess uncertainty about infinities, even of the denumerable kind (Barcan Marcus, 1993a [1961]: 27; 1978: 360–361). Both criteria have a substantial advantage in one regard, and a substantial disadvantage in the other; Barcan Marcus' tag criterion best accommodates direct reference theorists; Quine's quantificational criterion is preferable for proponents of objectual quantification over non-denumerable domains.

Acknowledgments

Thanks are due to audiences at the Universities of Campinas, Glasgow, McMaster, and Rome La Sapienza, and also to the people who attended my postgraduate seminar *Logical Form and Ontology* at the University of Campinas in 2014. I am also grateful to Gary Ebbs, Jane Heal, Fraser MacBride, and Peter Sullivan for discussion. This research was supported by a Capes Postdoctoral Research Fellowship Grant.

Notes

1. For further details see MacBride and Janssen-Lauret, forthcoming 2015, section 3.
2. This paper, dated 1939, remained unpublished until the first edition of *The Ways of Paradox* in 1966. The *Erkenntnis* volume it was due to appear in never materialized because of the outbreak of WWII.
3. Although this paper contains the phrase 'to be is to be the value of a variable' (Quine, 1939: 708), it also has Quine speaking of genuine names of things being substituends for variables.

4. The later Quine stopped calling himself a nominalist because he felt compelled to allow quantification over sets, insofar as it is indispensable to physics (Quine, 1981b). His views remained nominalistic in the sense of opposition to universals.
5. Barcan Marcus also enjoyed this pastime in a literal sense; see Hochberg (2014) for his account of her challenging him to a duel in a castle.
6. Quine continued to maintain the anti-essentialist conclusion that although it makes some sense, in context, to say that rationality is essential to mathematicians or two-leggedness to cyclists, it makes no sense to say of an individual, independently of how that individual is described, that he or she is necessarily rational or bipedal. This argument is also deeply interwoven with Quine's global holism, and with that in mind the conclusion is to some extent defensible. See MacBride and Janssen-Lauret (forthcoming 2015).
7. See also (Barcan Marcus, 1967).

Bibliography

R. Barcan (1947) 'The Identity of Individuals in a Strict Functional Calculus of Second Order', *The Journal of Symbolic Logic*, 12(1): 12–15.
R. Barcan Marcus (1961) 'Modalities and Intensional Languages', *Synthese*, 13(4): 302–322.
R. Barcan Marcus (1967) 'Essentialism in Modal Logic', *Noûs*, 1(1): 90–96.
R. Barcan Marcus (1972) 'Quantification and Ontology', *Noûs*, 5(3): 187–202.
R. Barcan Marcus (1978) 'Nominalism and the Substitutional Quantifier', *The Monist*, 61(3): 351–362.
R. Barcan Marcus (1990) 'A Backward Look at Quine's Animadversions on Modalities'. In R. Barrett and R. Gibson (eds), *Perspectives on Quine*. Blackwell, Oxford, pp. 230–243. Reprinted in *Modalities* (New York: Oxford University Press).
R. Barcan Marcus (1993a [1961]) 'Modalities and Intensional Languages Appendix 1A: Discussion', in *Modalities* (New York: Oxford University Press).
R. Barcan Marcus (1993b [1985]) 'Possibilia and Possible Worlds', in *Modalities* (New York: Oxford University Press).
Y. Ben-Menahem (2015) 'The Web and the Tree: Quine and James on the Growth of Knowledge', this volume.
R. Carnap (1950) 'Empricism, Semantics and Ontology', *Revue Internationale de Philosophie*, 4: 20–40.
R. Cartwright, (1954) 'Ontology and the Theory of Meaning.' *Philosophy of Science* 21: 316–325.
M. Dunn and N. Belnap (1968) 'The Substitution Interpretation of the Quantifiers', *Noûs*, 2(2): 177–185.
H. Hochberg (2014) 'Some Things Recalled', *Dialectica*, 68(2): 171–182.
T. Hofweber (2005) 'A Puzzle about Ontology', *Noûs*, 39: 256–283.
P. Horwich (1990) *Truth* (Oxford: Basil Blackwell).
S. Kripke (1980) *Naming and Necessity* (Cambridge: Harvard University Press).
E.J. Lowe (2006) *The Four-Category Ontology* (Oxford: Clarendon Press).

F. MacBride and F. Janssen-Lauret (forthcoming 2015) 'Meta-Ontology, Epistemology & Essence: On the Empirical Deduction of the Categories', *The Monist*, 98(3).

H. Price (2009) 'Metaphysics after Carnap: The ghost who walks?'. In D.J. Chalmers, D. Manley, and R. Wasserman (eds.), *Metametaphysics*. Oxford: Oxford University Press, pp. 320–346.

W.V. Quine (1939a) 'Designation and Existence', *The Journal of Philosophy*, 39: 701–709.

W.V. Quine (1943) 'Notes on Existence and Necessity', *The Journal of Philosophy*, 40: 113–127.

W.V. Quine (1944) *O Sentido da Nova Logica* (Sao Paulo: Martins).

W.V. Quine (1947a) 'On Universals', *The Journal of Symbolic Logic*, 12(3): 74–84.

W.V. Quine (1947b) 'Review of "The Identity of Individuals in a Strict Functional Calculus of Second Order" by Ruth C. Barcan', *The Journal of Symbolic Logic*, 12(3): 95–96.

W.V. Quine (1947c) 'The Problem of Interpreting Modal Logic', *The Journal of Symbolic Logic*, 12(2): 43–48.

W.V. Quine (1951a) 'On Carnap's Views on Ontology', *Philosophical Studies*, 2: 65–72.

W.V. Quine (1951b), 'Two Dogmas of Empiricism', *The Philosophical Review*, 60: 20–43.

W.V. Quine (1953) 'Reference and Modality'. In *From a Logical Point of View*. Cambridge, Mass: Harvard University Press, pp. 139–159.

W.V. Quine (1953 [1950]) 'Identity, Ostension, and Hypostasis'. In *From a Logical Point of View*. Cambridge, Mass: Harvard University Press.

W.V. Quine (1957–8) 'Speaking of Objects', *Proceedings and Addresses of the American Philosophical Association*, 31: 5–22.

W.V. Quine (1960) *Word and Object* (Cambridge, Mass: MIT Press).

W.V. Quine (1968) 'Ontological Relativity', *The Journal of Philosophy*, 65(7): 185–212.

W.V. Quine (1969) 'Existence and quantification', in *Ontological Relativity and Other Essays* (New York: Columbia University Press).

W.V. Quine (1970) *Philosophy of Logic* (Englewood Cliffs: Prentice-Hall).

W.V. Quine (1976a) 'Worlds Away', *The Journal of Philosophy*, 73(22): 859–863.

W.V. Quine (1976b [1939]) 'A Logistical Approach to the Ontological Problem', in *The Ways of Paradox*, second edn (Cambridge, Mass: Harvard University Press).

W.V. Quine (1976c [1954]) 'The Scope and Language of Science', in *The Ways of Paradox*, second edn (Cambridge, Mass: Harvard University Press).

W.V. Quine (1976d [1955]) 'Posits and Reality', in *The Ways of Paradox*, second edn (Cambridge, Mass: Harvard University Press).

W.V. Quine (1976e [1961]) 'Reply to Professor Marcus', in *The Ways of Paradox*, second edn (Cambridge, Mass: Harvard University Press).

W.V. Quine (1976f [1970]) 'Truth and Disquotation', in *The Ways of Paradox*, second edn (Cambridge, Mass: Harvard University Press).

W.V. Quine (1979) 'On not learning to quantify', *The Journal of Philosophy*, 76: 429–430.

W.V. Quine (1981a) 'The Pragmatists' Place in Empiricism'. In R.J. Mulvaney and P.M. Zeltner (eds.) 1981, *Pragmatism: Its Sources and Prospects*. Columbia: University of South Carolina Press.

W.V. Quine (1981b) 'Things and Their Place in Theories', in *Theories and Things,* (Cambridge, Mass: Harvard University Press).

W.V. Quine (1988) 'Comment on Agassi's remarks', *Zeitschrift für allgemeine Wissenschaftstheorie / Journal for General Philosophy of Science,* 19(2): 117–118.

W.V. Quine (1992) *Pursuit of Truth,* revised edn (Cambridge, Mass: Harvard University Press).

J. Schaffer (2009) 'On what grounds what'. In D.J. Chalmers, D. Manley, and R. Wasserman (eds.), *Metametaphysics.* Oxford: Oxford University Press.

S. Shapiro (2000) *Foundations without Foundationalism,* second edn (Oxford: Oxford University Press).

P.F. Strawon (1959) 'Individuals' (London: Methuen).

T. Tahko (2011) 'In defence of Aristotelian metaphysics', in *Contemporary Aristotelian Metaphysics* (Cambridge: Cambridge University Press).

A. Tarski, (1956 [1933]) 'The Concept of Truth in Formalised Languages', in *Logic, Semantics, Metamathematics* (Oxford: Clarendon Press).

P. van Inwagen (1998) 'Meta-ontology', *Erkenntnis,* 48: 233–250.

12
Underdetermination, Realism, and Transcendental Metaphysics in Quine

Gary Kemp

A certain variety of anti-realist thinks of accepted natural science, or significant portions of it, as not literally true, or as in some sense not measuring up to the standards for realism revealed by philosophers. By contrast, the naturalism that Quine so persistently espoused is by its own lights a species of scientific realism. It holds that there is no point of view that stands above science – the various sciences including mathematics and logic – from which one could gainsay or substantially re-interpret large swaths of its findings. It is naturalized epistemology, but also naturalized metaphysics, even if Quine did not call it that.[1] Quine is explicit that the very ideas of truth, reference, objectivity and so on have only the sense afforded to them within natural science itself (or rather: *regimented* natural science – roughly, the most streamlined version of it as represented in the first-order predicate calculus with identity). Realism about the external world, about the past and future, and about induction, to take the most general examples, are by and large assumed by science (and by common sense); there is no other point of view, no higher standard, no further question with respect to such commitments. To be sure, they are factual assumptions that could conceivably be overturned, but are as well-founded as any part of the general naturalistic world view.

In what follows I consider an apparent challenge for Quine's realism, and by implication to his naturalism, that issues from what is often thought to be another longstanding commitment of his: the Underdetermination of Theory. The thesis was very much in the air at the time when Quine's philosophy came to be shaped in the 1930s and 1940s; many philosophers of the Vienna Circle and those clustered vaguely around the Unity of Science movement – Hans Hahn (1879–1934), Otto Neurath (1882–1945), Philipp Frank (1884–1966), Rudolf Carnap (1891–1970), Charles Morris (1901–1979), and Herbert Feigl (1902–1988), being the major figures – were well known to Quine, and felt the influence of the idea and contributed to its explication. These figures were influenced in turn by earlier Logical Positivists including Ernst Mach (1838–1916) and Morris Schlick (1882–1936), and by the conventionalist line of such figures as Pierre Duhem (1861–1916) and Henri Poincare (1854–1912). Albert Einstein (1879–1955) himself, who was known to Quine somewhat later, also espoused the doctrine (Howard, 1990). Later such influential figures in the philosophy of science as Thomas Kuhn (1922–1996), Paul Feyerabend (1924–1994), and Imre Lakatos (1922–1974) accepted the idea in one form or another. More recently the idea is vital to the Constructive Empiricism of Bas van Fraassen in his *The Scientific Image* (1980) and after. For van Fraassen it is a central reason to hold that the aim of science beyond observation is not literal truth but empirical adequacy.

More generally the thesis is often felt to sit uneasily with realism. If, for example, the nucleus in a helium atom contains two protons and two neutrons, then surely a theory which had it otherwise could not be true, or if true then it would still come up short on the score of its realism. Quine's holism of 'Two Dogmas of Empiricism' of 1951 is often thought straightforwardly to imply underdetermination (and therefore, to some, some kind of anti-realism, perhaps instrumentalism). But as some astute commentators know – for example Gibson, Hylton, Ben-Menahem, and Severo – Quine's attitude towards the thesis changed as his epistemological views became subtler and more refined; it changed more than once, and sometimes changed in response to his readers and critics (Gibson, 1991). There is more complexity and nuance than meets the eye, both in the exact formulation of the thesis and in its role and significance in Quine's overall philosophy. For example, if in 1951 he did think that the thesis simply follows from his holism, by the time of *Word and Object* of 1960, he thought it significant that it could be viewed as being established by – not, as he would for a time later, as establishing – the thesis of the indeterminacy of translation, the argument for which is for many people less than pellucid (1960: 78). And when he eventually devoted a full article to the topic – 'Empirically Equivalent Systems of the World' of 1975 – he professes ambivalence: 'The doctrine', he writes of the underdetermination of theory, 'is plausible insofar as it is intelligible, but it is less readily intelligible that it may seem' (Quine, 2008a[1975]:

228). *Pursuit of Truth* (revised edition of 1991) saw further changes, as we'll see.

On balance, I think that Quine's mature views do question the intelligibility of the thesis, and therefore to that extent the apparent tension – between realism and the underdetermination of theory – is to that extent resolved. I'll first explain what the thesis of underdetermination comes to in Quine's naturalistic picture. Underdetermination appears in the general literature in various guises: Some are threatening, some are trivial; some are epistemological or methodological, some are factual or metaphysical. After isolating the sense which is relevant to Quine's philosophy, I will try to evaluate whether or in what way the underdetermination of theory is genuinely a problem for the Quinean naturalist, and ultimately cast doubt on whether underdetermination really makes enough naturalistic sense not to be discounted.

1   Holism and the thesis of underdetermination

The idea of underdetermination has in one form or another been around for at least as long as the conversation about skepticism in the seventeenth century. Sensory experience, said Descartes, cannot choose between the hypothesis of a material world and that of an Evil Genius manipulating one's perceptions just so. He thought there was a reasonable way to choose, but others were less convinced. Hume is often read as one who accepts the skeptical conclusion of such reasonings: that we have no reason to prefer one competing explanation of our experience over another, and thus all that we really know is experience itself.

The discussion took a more precise form and became pressing not only for philosophy or the philosophy of science, but for science itself, and in particular for physics, at the time of the careful and historically detailed work of Duhem in his *The Aim and Structure of Physical Theory* of 1906. The basic idea can be grasped with a simple example. We are told (we 'observe') that the sum of x and y is 12; what are x and y? Eight and four? Ten and two? Or, 3,011 and negative 2,999? There are infinitely many possible correct answers – many 'explanations' or 'theories' for the 'observations' or 'data'. According to Duhem, it is inevitable that the same situation – multiple competing explanations for the data we are given – holds for theories within physics, even though the range of data is comparatively vast, and the explanations or theories much more complex.

In more detail: Suppose a physical theory A comprises a finite set of theoretical hypotheses and other principles $P_1..P_k$, which implies some infinite set of observations $o_1 \ldots o_n \ldots$, only some of which have actually been observed. Suppose now that *not-o_k* is observed, conflicting with the implication of A that o_k. Rationality demands that one should seek to adjust $P_1..P_k$ to erase the conflict. But the conflict doesn't tell which one of $P_1..P_k$ to change. Even if one is testing a certain hypothesis that leads to the expectation of o_k, and the result is not-o_k, one could, practicalities aside, keep the hypothesis, and adjust other regions of the theory so as not to imply o_k, or to imply not-o_k. Normally one does not think of the other regions as up for testing, but the point is just that one could.

Thus Duhem's thesis of confirmational holism: Strictly speaking, only large chunks of physical theory, and not isolated theoretical statements, are susceptible to testing (Duhem, 1954 [1906]: 180–200). The thesis should be understood as counterfactual-supporting: Even statements of theories or stretches of theory that are in fact well borne out by observation are still supported only holistically by those observations, for if some observational implication had been falsified – contrary to fact – then there would remain the same latitude for revision of the theory.

That much might seem almost trivial, at least to an empiricist. More substance accrues to the idea when one reflects that these decisions will often have implications for other parts of science, and these others with others, and so on. In principle, for mid-period Quine at any rate, it is not particular theoretical statements that are up for testing in a given case, not even large chunks of science, but the whole of science.[2] Thus the maximal version of the thesis of confirmational holism (the maximal version was made famous by Quine's 'Two Dogmas of Empiricism' in Quine, 1961: 42–46; Duhem articulated only the smaller-scale version (1954 [1906]:180–200, especially 187).[3]

How do we get from confirmational holism to underdetermination? Quine observes that science in not monolithic; it is loose in its joints in various degrees (2008a[1975]: 230). Nevertheless he holds we can speak of a single system of nature not only because of the various parts being woven together by logic and mathematics, but also because the various special sciences, even if not strictly speaking reducible to physics, supervene at least vaguely on that fundamental science. Thus, if one can find multiple sets of materially different theoretical hypotheses which are equivalent with respect to a certain critical range of observation, then it is by no means obvious how one might rule out that the point holds when the range is expanded to *all* observations, and thus to the entire system of theoretical statements whose business is to imply them. Thus we have the thesis of the fully-fledged underdetermination of theory: any theory – including the entirety of our general theory of nature – is in principle observationally equivalent to some other significantly different theory (Quine, 2008a [1975]: 228–229).

Still one may accept maximal confirmational holism while rejecting underdetermination, since one can accept that when one observes or conducts experiments, in principle one is always testing the whole of science, without thinking that there must be an observationally equivalent rival; perhaps the bump would re-appear elsewhere in the rug. And surely the other way round does not hold either: Quine himself does not, by the time of writing at length on the topic in 1975, accept maximal holism, but he does accept underdetermination. Perhaps there is no tight argument for the latter from the plausible thesis of a more moderate holism, but for Quine, the possibility cannot be ruled out.

It is important to emphasise that the thesis of underdetermination of theory involves much more than just the fact that a theory logically implies its data but the data don't logically imply the theory: the thesis is that substantially different overall theories can imply the *same* observations, that they are empirically equivalent. It is also worth stressing again that it is a global thesis. The idea is that two theories might be evidentially equivalent theories of the spatio-temporal whole of nature. They are not merely such that no evidence we have decides between them, but are such that *no* evidence – past, present, or future – *could* decide between them: underdetermination of total science concerning all possible observations.

More precise content of the thesis awaits explanations of observation and theory, and of the distinction between them. And this is where the distinctiveness of Quine's position begins to emerge. To understand this, we have to understand in outline Quine's naturalism, and in particular his manner of schematising language. In doing so, we will find that for Quine, the content of the thesis is indeed inseparable from his philosophy of language, and its significance and interest are bound up with his philosophy generally.

2   Details of Quine's version

Quine looks on theories and their evidence as being matters of language – in particular the declarative sentences of language (for our purposes we can assume that the theories in view are housed in the same language).[4] He divides the realm of declarative sentences into the occasion sentences and the standing sentences. As opposed to standing sentences, occasion sentences are those whose truth-values vary with time. A crucial subset of these latter are *observation sentences*. Observation sentences are occasion sentences to which a given speaker is disposed to assent just in case some set of the speaker's exteroceptors – the sensory nerves that are sensitive to events outside or at the surface of the body – are activated, and similarly for dissent (non-observational occasion sentences have no such correlation). They are also such that any other member of the linguistic community would agree with the subject's verdict if he were to undergo a similar stimulation (as time went on, Quine relaxed this requirement, for it presupposes something that ought not to matter, namely that the relevant subjects are neurologically similar; this is an important point but it will not figure in what follows; see 1969: 157–160; 1992: 40–44; 1995: 19–21; also see Kemp, 2012: 32–33, 128–143; and also see in this volume 'Pre-established Harmony' by Quine and 'Introduction to Two Previously Unpublished Articles' by Gary Ebbs).

Other declarative sentences include *theoretical sentences*, all of which are standing sentences. The simplest type of theoretical sentence is the observation categorical, such as 'If smoke, fire'; these – not observation sentences themselves – are what a scientist tests; the scientist contrives or looks for the truth of the antecedent, and checks for the truth-value of the consequent.[5] Theoretical sentences which are not observation categoricals – 'RNA is less stable than DNA', '$E=mc^2$' and so on – are much more complicated, but they are ultimately sensitive to observation in that they imply, more circuitously and together with many other theoretical sentences, the relevant observation categoricals.

By a theory's *empirical content* Quine means the sum of its observation categoricals (Quine, 1981: 28; 1992: 16–18).

The immediate evidence for a speaker's theory is not the sensory stimulations themselves – and not the objects and events perceived and not sense-data – but a certain subset of the observation sentences of the subject's language such that each is paired with an actual context in which the subject has or had the disposition to assent to it or dissent from it (Quine, 1969: 69–90; 2008a[1993]: 413; 2008a[1975]: 230; 1992: 1–18; 1995: 16–26; cf. the notion of a 'protocol sentence' in Schlick 1959[1932] and Neurath 1959[1934]; and for more detail see Johnsen 2014 and Sinclair 2014).[6] The effect is thus a variety of standing sentence (an 'eternal sentence'; closely related is the notion of a 'pegged observation sentence', which is any observation sentence irrespective of whether or not any speaker made the relevant observation, together with temporal and spatial coordinates 2008a[1975]: 232; and see below). The stimulation perchance causes the disposition to assent to an observation sentence, but is not in general part of the evidence or the empirical content of a theory (Quine, 1981: 40). The evidence delivered by an utterance 'It's warm' is simply that it's warm for a certain speaker at certain time and place: an observation sentence plus the relevant matters of context.

This is *not* meant as an analysis of the ordinary concept of evidence or of that of observation, at least partly because of the imprecise and fluctuating character of those ordinary concepts, the various uses of the words. It's meant to show that questions about the relation of evidence or observation to theory can be made satisfactory sense of by recourse to the official Quinean story of observation sentences and their relation to the rest of theory (Quine, 1992: 2; 2008a[1975]: 231); the strategy is very much in the spirit of Carnapian explication. In a passage from *Pursuit of Truth* he went so far as it explicitly expunge the words 'observation' and 'evidence' from the story:

> Observation then drops out as a technical notion. So does evidence, if that was observation. We can deal with the question of evidence for science without help of "evidence" as a technical term. We can make do instead with the notion of observation sentence. (1992: 2; see also 1974: 38–39)

Since, for linguistically competent individuals at any rate, it stands to reason that one perceives or observes warmth just in case one becomes disposed to assent to 'It's warm' (or some translation of it), we can say that the ordinary notion of one's evidence as what one perceives or observes, as something suitable for expression by uttering observation sentences, can be replaced in Quine's scheme by the observation sentences themselves (along with identity of time and speaker).

The observation categoricals that constitute the empirical content of a theory appertain to many more circumstances than those at which a given observer is present or those at which some observer or other is present. In fact the underdetermination thesis pertains to *all* conditionals comprising pegged observation sentences as antecedent and consequent, not just those which count as a given speaker's evidence in the sense just described; thus it concerns – it is an idealisation of course – all observation sentences pegged to every region of space-time however inaccessible.[7] Sometimes I will continue to speak loosely of evidence, but officially I mean it in the sense of empirical content – observation categoricals, covering all pegged observation sentences.

So far I have spoken only there being empirically equivalent but 'different' global theories of nature, without indicating further what this difference amounts to. When Quine first discussed at length the idea of underdetermination, he spoke initially of outright logical incompatibility: Theory B might imply the negation of a sentence implied by an empirically equivalent theory A (2008a [1975]: 237). However Quine, at the suggestion of Donald Davidson, came to think that any such logical incompatibility can repaired along the following lines. If A implies, for example, that neutrinos have mass, but B implies that neutrinos do not have mass, one can simply say that it only appears to be a contradiction, that there is an equivocation in the two uses of the syntactical shape 'neutrino'; one should speak rather of 'neutrinos$_A$' and 'neutrinos$_B$'. We can repeat the procedure if necessary, until the apparent contradictions disappear (1992: 97–98; for more see 2008a [1975]: 242 and 2008a [1986]: 335–337).

Quine came to view the envisaged incompatibility as not logical but practical, perhaps even as psychological: A and B have the same empirical content, but they cannot, despite our best efforts, be reconciled (2008a[1975]: 242–243; 1992: 97). In particular, for at least one theoretical sentence of A, we cannot find a theoretical sentence or conjunction of such sentences of B that can serve in lieu of the A-sentence, or rather which plays the same inferential role in B that the theoretical sentence played in A. The difference would have to be due to at least one theoretical term that is essential to the one theory but absent from the other, and neither can one define it using expressions of the other. (1992: 96–97).

Any astute reader of 'Two Dogmas of Empiricism' will want to know what has become of the other scientific virtues that Quine mentions in the essay, as well as ones he described in later work, such as simplicity, familiarity, scope, conservatism, fecundity, clarity, capacity for use in prediction and technology, and more generally and vaguely for understanding nature (see especially *The Web of Belief*, Quine and Ullian, 1978). Quine readily admits that the schematism of empirical content and observation categoricals is an idealisation of actual scientific practice, that his 'concern [is merely] with the central logical structure of empirical evidence' (1992: 18). The schematism applies straightforwardly only to 'testable' theories or sets of sentences (1992: 18). If we speak of these other virtues together with empirical contentfulness as measuring a theory's 'scientific value' (1992: 95–96), then, in comparison with empirical content itself, the scientific value of a theory is obviously a multifaceted and relatively vague matter. As is well-described by Lars Bergström, it is inevitable that subjective factors will creep in (Bergström, 1993:98–100). We could speak of 'scientific equivalence' of theories, allowing for all these virtues, but I will restrict the discussion to the empirical equivalence of (testable) theories, leaving aside comparisons of their overall scientific value.

Last but certainly not least comes the question of truth. If A and B were total theories of nature with the same empirical content, we can allow that A and B would be equally *warranted*, but would A and B be equally *true*? Quine calls the response of saying they're both true an 'ecumenical' one, and saying that only one is true a 'sectarian' one (Quine, 1992: 98–101). The ecumenicalist position, which Quine affirmed in 1975, involves accepting both theories as one giant theory (though he was thinking then of logically incompatible theories, thus envisaging a two-sorted truth-predicate). But this acceptance, he came to decide, is gratuitous – all that added theory without a whit of added coverage of observables. Obeying Ockam's razor, the sectarian by contrast settles for a frank dualism: one is free to speak in terms of one or the other, but not both simultaneously, and is bound by the principles of the theory one is using. We are still free to shift between them for whatever specific reason that happens to arise:

> he is as free as the ecumenist to oscillate between the two theories for the sake of added perspective from which to triangulate on problems. In his sectarian way he does deem the one theory true and the alien terms of the other theory meaningless, but only so long as he is entertaining the one theory rather than the other. He can readily shift the shoe to the other foot. (1992: 100)

Before getting on to the next topic, we should stress certain consequences of looking at this issue from the standpoint of Quine's naturalism. For the Quinean naturalist, components of 'our theory' – in particular those components describing evidence and theory as matters of language – are invoked in expounding the naturalistic view of language. In particular, where total theories of nature are at issue, warrant, like every other theoretical measure, has strictly to be internal to the theory – has to be something generated from within it, has to be immanent rather than transcendent. It's a circle, but a benign one. The underdetermination thesis is thus not an *a priori*, theoretically neutral idea, but one whose precise content draws on our overall theory of nature.

And the position is striking that a theory's not being uniquely warranted by the evidence does *not* impugn that degree of warrant (and that Quine does not recognise, in addition to the sectarian and ecumenicalist responses to the prospect of underdetermination, a nihilist one, which refuses to assert either A or B on account neither's being best). It is part and parcel of Quinean naturalism that the underdetermination thesis does not present an epistemological challenge, a skeptical threat. For Quine, close analysis of Carnap's failure to show how theories of the external world can be reduced to a basis in sensory experience plus logic did not raise the spectre of skepticism; Carnap's interest like Quine's was to explicate the logical structure of science, not to justify science. But that failure did show that philosophy must begin in *medias res*, in the middle of things, with one taking the deliverances of the natural sciences for true (Quine, 1961: 39–40). Not only do we have to 'stop dreaming of deducing science from observations' (Quine, 1969: 76), we must despair of '*translating* the sentences of sentences of science into terms of observation, logic, and set theory' (1969: 77; emphasis in the original). Nor does it achieve anything substantial to count the sentences which link observation to theory 'reduction forms' (1969: 76); that would be a clear case of mere word-making. At the same time, one cannot assume the negation of the underdetermination thesis. That is the most general reason that any question raised by the thesis of underdetermination is not, for Quine, an epistemological one: if it raises a question, then it is a metaphysical one, about what reality is.

3   Worries over Quine's position

So was Quine really a realist? The word does not figure prominently in his writings, but he evidently thought so. According to his naturalism, there is no standpoint besides the scientific standpoint to judge whether certain entities are real, whether they exist, whether the procedures that issue in our theory of such entities are fully objective, how such entities stand to one another, and so on. In 'Posits and Reality' (Quine, 1976: 246–254), in a point that strongly parallels both Austin and Wittgenstein (and also, if less strongly, Moore, Carnap, and the early Ayer), even commonsense judgements involving such concepts as *existence* and *reality* – the concepts of intuitive metaphysics – do not somehow reach beyond the significance that is afforded to them in ordinary language training (Quine, 1976: 251–254; also 1963 [1951]: 44; 1960: 22; 2008b: 152; 2008a: 405). One learns to call narwhals real, unicorns unreal, and that is that. There is no conflict between the reality of material objects and the naturalised epistemologist's concluding that they, along with *all* objects, are 'posits': it just means that our use of such ordinary expressions as 'the moon' is, from the epistemologist's reconstructive point of view, optional. Quine regards 'all objects as theoretical' (1980: 20). So if our overall theory is A, then to assert the sentences of A is to commit to the ontology of A, to the truth of A, and to A as telling us of reality. It might be a flat-footed realism, not the transcendental realism hankered after by some people; it might even strike one as cheat, or as opening the door to relativism, but according to Quine it is realism all the same, in the only sense there is that he recognises.[8]

But it's hard to be altogether satisfied by this. If some different way of speaking, some hypothetical theory of the universe, would have precisely the same empirical warrant and claim to truth as ours, but not join with ours in asserting the reality of black holes or the facts of evolution in the sense we mean them, then is there not something lacking in advertising the position as one of realism? Surely to be a realist is to believe, contrary to Quine, that there *is* just 'one solution to the riddle of the universe' (2008a: 270). Quine may have considered the thesis as debateable, may have 'vacillated' (1991: 101) over it, but definitely did not consider its being an open question to be a threat to his realism, or as reason to concede that his position is one of realism but only in his limited sense. I'm going to (1) try to reduce the scope of underdetermination within Quine's naturalism, and then (2), in response to a point of Adrian Moore's, try to make it palatable that there is there just isn't a higher standard for realism than what Quine describes. This suggests, not merely a reiteration of naturalism, but also a concern with the significance or importance of the thesis.

3.1   Within our theory: A further naturalistic constraint on underdetermination

According to what scientists say, we human beings can only sense a small portion of the forces impinging upon us. In the first place, within a given phenomenon that we can sense, we're sensitive to only a limited range within that phenomenon (and within that range, our capacity for fineness of discrimination is very limited). Human vision detects only a miniscule band of electromagnetic radiation. The band we call 'light' is sandwiched between radio waves, microwaves, and infra-red radiation on the one hand, and ultraviolet 'light', x-rays, gamma rays, and cosmic rays on the other; it is all photon-emission, all the same kind of phenomenon, just different in frequency. The waves in the air and other media that we are familiar with as 'sound' go much higher and a bit lower in frequency than ones that we can sense, as we know from whales (whose hearing goes well under 20hz) and dogs (whose hearing goes well over 20,000hz). In the second place, there are phenomena which we have no means at all of sensing: radioactivity, magnetic and electric fields, and positron emission on the one hand, quantum forces on the other. The capacity for electroreception in platypuses is an example. And third, the animal kingdom is full of clever ways of detecting objects and features of the environment which perhaps do not strictly require an additional sense, but which do require cognitive apparatuses or enhancement of sensory refinement which we lack. Nagel famously pointed out the capacity for echolocation in bats.

Some of these forces are not sensible by any creature on earth or elsewhere, but there is no reason not to speak cheerfully of senses which go completely unrealised, so long as such a sense can be given a third-person description in terms of what we know of nature: eyes like electronic microscopes, or giant telescopes, or which are sensitive to other regions of the electromagnetic spectrum like radio or gamma-rays; or the ability to sense directly magnetism, radioactivity, the emission of sub-atomic particles, and so on. Now according to Quine's definition of empirical equivalence, theories A and B are empirically equivalent if and only if the set of observation categoricals implied by A and B is the same. And observation categoricals comprise observation sentences together with 'If-then' (roughly). This makes the empirical content of a theory relative to what counts as observation, on what sensory apparatuses are in view; A and B may be empirically equivalent for us yet not for creatures with different senses.

Now imagine a super-creature, fitted with all possible senses (or: every sense that we can coherently describe). For such a creature, far less will be theoretical, far more will be observational. The ultimate such creature is an ideal posit of reason, we might say: a *sensus optimus*. It is a mirror image of Kant's idea of a pure intellect. The ultimate super-being is presumably impossible, and it is not feasible to give precise reasons for the place on the scale towards it beyond which there is no possible being. But progress up the scale towards the limit is obviously conceivable, and that is enough.

In more detail. Within physics and biology as presently constituted, we can imagine different constellations of senses. In particular, we can imagine indefinitely many creatures C1, C2... with entirely different senses from ours, sharing none of ours, but which in a certain rough sense are equally well-placed with respect to reality as we are; their class of observation sentences would be disjoint from ours, but play the same role in their language as ours do in ours (I will not stop to clarify this notion of 'same role', but I assume it makes sense intuitively). In fact all of the creatures C1, C2... can share in a certain sense the same theory, in that they consist of exactly the same sentences with the same truth-values. For suppose that our present science is the theory A, with

E a subset of A, its observation categoricals. And suppose there are in A indefinitely many disjoint analogues to that empirical content, E1, E2...(corresponding to the creatures C1, C2...) – all nonetheless in service of A – where a sentence is a member of an evidence-stock only if it is not a member of another (so such a sentence as 'If a substance is magnetically charged, then iron is attracted', which is not strictly an observation categorical for us, is still a member of A). Richer sets of observation categoricals can be conceived which do overlap with others, with the culmination reached at the sensus optimus (still in service of A). What counts as evidence is still relative to the creature. As we consider creatures further up the slope of sensory sensitivity, with more and more sensory possibilities realised, we find the range of possibilities constrained, converging on the sensus optimus. Yet the theory of the sensus optimus remains A.

We cannot discount the underdetermination of theory even for a creature at or near the limit, but underdetermination for the creature would have far less scope. If we can speak of the degree of the inverse of underdetermination as the degree of determination, then the degree of determination will track the place of a creature's theory on the scale. As the range of observation sentences expands, as the range of empirical checkpoints broadens, the scope of underdetermination contracts.

Quinean naturalism, then, has the resources for reducing the slack between theory and observation. The tightening is only conceivable, not actual, but it passes the test of naturalism. Indeed the character of our present theory supports that thesis: An occasion sentence such as 'The substance emits an alpha-particle' – which again is not an observation sentence for us, but which would be an observation sentence for a creature with the sensory analogue of a Geiger counter – supports the thesis that our theory admits of such tightening. The most reasonable conjecture on the basis of available evidence is simply that the actual scope of indeterminacy would decrease if our senses where sharper and more various.

Not that the potential tightening is potential elimination in any sense; the sensory optimus has its limits. An unlimited sensory optimus would be involve the transformation of all theoretical sentences on the part of a creature who literally saw everything, from the big bang to the final end in the minutest detail to the grandest views, into observation categoricals; this is not consistent with what we know of knowing beings (and still such a creature presumably would require other standing sentences – at least sentences of mathematics and logic – to stitch its theory together). According to Quinean naturalism, the underdetermination of theory is an inescapable fact of life.

3.2 Without our theory: naturalism and metaphysics

We have just considered cases of the same theory A served by differing but not incompatible sets of evidence. The thought experiment put some pressure on the full-blown thesis of whole-sale underdetermination. Now we directly consider again the spectre of global underdetermination in principle, the same set of evidence serving different theories. A way into the problem that I will discuss has recently been articulated by Adrian Moore:

> Quine's lax sectarianism, which is laxer than his naturalism warrants, exhibits a dim recognition that only something less extreme is ultimately sustainable. For in his lax sectarianism Quine allows himself to step back from the scientific way of making sense of things, which is the only way of making sense of things that the extreme view sanctions, and tries to make sense of that way of making sense of things without simply redeploying it. (2012: 324)

Consider the predicament of someone who is confronted with A and B. According to Quine's naturalism, one is in the situation of the Neurathean sailor: all knowledge including philosophy is in the same boat – say boat A – and the idea that one could rise above one's boat, or jump off as Moore elsewhere suggests, is illusory. But it strains credulity to think that A and B are *absolutely incommensurable*. Quine appears to admit as much:

> the rival theories describe one and the same world. Limited to our human terms and devices, we grasp the world variously. I think of the disparate ways of getting at the diameter of an impenetrable sphere: we may pinion the sphere in calipers or we may girdle it with a tape measure and divide by pi, but there is no getting inside. (1992: 101).

Quine is not imagining schizophrenia. In order to make sense of this transcendental geometer, or the rogue Neurathean sailor, it seems we have to grant him or her some conceptual equipage – irrespective of which theory he chooses, to make sense of himself as rationally considering the choice. As we noted, Quine insists that words like 'reality' and 'truth,' and presumably phrases such as 'the world', are meaningful only thanks to the sort of contexts in which they are learned. But surely such vocabulary does not change significantly in its import depending on what total theory of the universe one has in view. We can ask for the *ontology* of a theory, for what it takes to be *real*, without supposing that the answers will somehow change the sense of the question out of all recognition. A and B are indeed competing theories of the *same world* (presumably there is only one), differing in the entities they take to be real, how those entities combine, and so forth. The comparison itself is a well-formed question: does A or B, or both, or neither, tell us of reality? We must admit that knowing beings must have limits, but the limits are limits to the knower's relation to reality, to the world, which is therefore something known to exist but which we can know only very incompletely. This is something we know independently of the particular scientific theory we happen to espouse. It is knowledge for which Quine's rigorous naturalism appears to be unable to find a place.

This is a somewhat familiar predicament in philosophy. A theory – say A – is presented as true, or as absolute, or absolute-for-us, or absolute-for-us-at-a-time, or absolute-for-us-at-a-time-with-due-allowances-for-human-shortcomings; then we wonder, 'What is the status of this judgement?'. One might suppose that there is an hierarchy of such judgements, such that the status of the judgement that (for example) *A is absolutely true* is itself subject to a further meta-level of scrutiny, the answer generated by this level is subject to a further meta-level of scrutiny, and so on. Metaphysical pronouncements are merely relative to their position in this Russell-like or Tarski-like hierarchy. But what of that statement? It seems for all the world as if *that* judgement extends across the entire hierarchy. So one concludes, 'Ah, so a bit of off-the-cuff reasoning shows that it is inconceivable that metaphysics is not absolute'.

Surely it can't be as easy as that. The trouble, as I think Quine would see it, is a certain parting of moorings, of language going on holiday. There are two main reasons why. On the one hand, if transcendental metaphysics is to pass muster theoretically, then some consistent explication has to be given of its key concepts: truth, reference, object, existence, and so on. For Quine, that is a central moral of the great story that runs through Zeno, Cantor, Russell, Gödel, and Tarski (e.g. 'The Ways of Paradox', in Quine, 1976: 1–18). Intuitive ideas in this realm are more or less bankrupt; to accept them as they are is to plunge back into metaphysics in the bad sense, like unteachable rats released once again into the labyrinth. The right response is carefully to examine their legitimate roles, and if necessary to modify the concepts or to construct surrogates. On the other hand, Quine's detailed programme – set in motion by 'Two Dogmas of Empiricism', and which was realised in maturity in *Word and Object*, *The Roots of Reference* and *From Stimulus to Science* – makes it hard to see the concepts of an *object* and of *reference* as genuinely enjoying lives that are independent of the genealogy of how we do come to employ ordinary nouns and form generalizations. In other words, as Peter Hylton puts it, reference, and the idea of an object, are theoretical through-and-through (Hylton, 2007: 304; see also 1993: 115–150). It is only by seeing in detail – if highly schematised detail – how the language-learning child comes gradually to acquire a mastery of referential language, do we see exactly what our blithe talk of 'about-ness' and 'object' really come to (Quine, 1980: 1–20).

So I do not think the Quinean is forced to acquiesce in transcendental metaphysics. For the Quinean, the subject is too vague, has too many options as to its point or purpose, to make anything but false edifices – an attitude familiar from Kant, as well as from Carnap and from Austin as described above. Its claims, questions, and doctrines involve cutting off too much of the life-blood of its main concepts. The Quinean should rather look upon the images of the transcendental geometer and the displaced Neurathean sailor as just that, images, or metaphors, optional ones that one is free to indulge in as one bumps up against the limits of language. The images do something, but carry little conceptual weight.

But not none. Again two points. First, as repeated above, a Quinean cannot *absolutely rule out* the possibility of an alternative theory of nature – there remains the sheer logical fact that our theory implies its data but not its data the theory, and it seems we cannot rule out that a different theory would imply the same data as ours.

Second, Quine considered, with respect to other subjects, a 'Doctrine of Gradualism': of degrees of analyticity, and of the relative theoreticity of observation sentences (1986: 100; 1971). And another area in which he considered a doctrine of gradualism – without naming it so – is in 'Facts of the Matter' where he speaks of 'ordinary language [as] only loosely factual' (2005: 285; see also 'Scope of the Language of Science', in Quine, 1976: 228–245), implying that the factuality or objectivity of a statement is not all-or-nothing, that it is a matter of degree. Ordinary thought and language are from a certain philosophical point of view notably flexible, porous, and protean, serving various purposes besides the rigorous statement of facts; and in any case they are only as rigid or precise as ordinary purposes demand.[9] Scientific and mathematical language, on the other hand, and especially if logically regimented as demanded by the Quinean ontological accountant, represent the maximum of a continuum of rigour and factuality, or as highly placed on the continuum as is humanly feasible. So in particular, the metaphysician can use the relatively rough versions of the concepts of metaphysics, making for relatively rough questions and doctrines. But any conclusions must be at most speculative, and will often be open to objections using the same rough tools. Such doctrines may be highly and fruitfully suggestive, they may serve usefully as scaffolding, even if they can never be the building itself.

But that is all. I doubt whether wholesale underdetermination really is sufficiently rigorous for the purpose of fuelling the transcendental metaphysicist's fire. The particular examples Quine cites are not fully-fledged global alternatives – they are not cases of what we might call *radical* underdetermination (Severo, 2008). Poincaré's comparison of normal centerless Newtonian space with the hypothesis of a centred space with sizes of objects decreasing at some constant rate as they move away from the centre are, let us assume, observationally equivalent – but call for relatively minor changes in the overall scheme (in fact for other reasons it does not strictly qualify as a case of underdetermination, as Quine admits; 2009[1975]: 237). Same thing with van Fraassen's example of a 'stationary' Newtonian universe versus one that moves as a whole in a certain direction at a constant rate. There remains considerable overlap with a standard Newtonian theory: Sentences of mathematics, chemistry, and biology would not be affected, for example, and physics itself would still retain much of the substance of its theories. It does not require the standpoint of the rogue Neurathean sailor to make sense of it.[10] Only the *extreme* case – the radical case – where much or all of a theory's theoretical vocabulary is irreconcilable with another's, are we forced to go transcendental, and thus, by naturalism, flirt with unintelligibility. Thus the possibility of radical or wholesale underdetermination should simply be discounted by the Quinean naturalist: logically speaking, it cannot be ruled out, but it asks one to consider a completely different scheme of nature which is nonetheless not described – and therefore itself looks too much like transcendental metaphysics, or even like the idle dreams of a philosopher. Somewhat like the other doctrines of transcendental metaphysics, it represents the limit to science, or scientific intelligibility or cognitive propriety. We may still consider A and B as Duhem did, where the underdetermination is not radical – where there is a sufficient degree of theoretical overlap – without indulging in transcendental metaphysics.[11]

We saw that according to the later Quine, the difference involved between two empirically equivalent theories has ultimately to be conceived as a mere pragmatic difference – that try as we might, we cannot find a theoretical statement or a group of theoretical statements of the one theory that are empirically equivalent to one or a group from the other theory. It is a case where what is said to be practically impossible is also not positively conceivable in any detail. Like the possibility that one is disembodied spirit being deceived by an evil genius, it's hard to feel the possibility as a real one, as threatening to our knowledge of our place in the universe.

Notes

1. Hylton 2007 uses the term as part of a discussion of Quine's views of coun- terfactuals, identity, attributes (properties and universals), abstract objects, causation, intensionality, extensionality, dispositions, objecthood, and of course ontology.
2. Care should be taken to distinguish confirmational holism from *semantic* holism, or meaning-holism, which is the thesis that the meaning of a word or sentence depends on that of the other words or sentences of the language. Quine does *not* accept semantic holism – or anti-holism – because he does not accept that any correct scientific theory treats of meanings of words or sentences (except the *stimulus meaning* of observation sentences, a very limited notion which does not begin to capture the intuitive notion of meaning).
3. The 'mid-period' caveat is due to Quine's having said, in 'Empirically Equivalent Systems of the World' (2008[1975]: 229–230) and 'Two Dogmas in Retrospect' (2008[1991]: 393), that radical holism is only 'legalistically' true. I believe that there is no straightforward route from modest holism to (radical) underdetermination.
4. The importance of Quine's focus on sentences also emerges in his conception of the identity of theories. Quine begins with a very tight notion of theory *formulations*: Theory formulations are identical if and only if they consist of the same words in the same logical structure. But not only do we want phys- ics-in-English and physics-in-German to count as the same theory, we want, for example, to allow that two statements of a theory in our language that are identical except for swapping the roles of the word 'electron' and 'molecule' as statements of the same theory; in fact we would allow this even if one of the two lacked a simple word for this. We can assume for our purposes that the first case is taken care of by accepting that smooth, unproblematic translations of a theory into different languages still state the same theory. For the second, Quine says that 'two formulations express the same theory if they are empiri- cally equivalent and there is a reconstrual of predicates that transforms the one theory into a logical equivalent of the other' (Quine, 2008[1975]: 235). Reconstruing an n-place predicate means supplying from the other theory an open sentence with n variables and the same extension. Thus theories can be identified with certain equivalence classes of formulations (236).

5. Actually the story of observation categoricals is more complicated. Strictly they are *not*, in themselves, ontologically committing, for the same reason that observation sentences are not: reference to objects is not required to explain them (reification emerges only with quantifiers). Psychologically, 'If smoke, fire' etc. express conditional expectations, our 'first faltering scientific laws', Quine calls them (Quine, 1995: 25); see 1992: 9–13. Quantification emerges with what Quine calls the *focussed observation categorical*, such as 'If there is a raven, then it is black' (as contrasted with the unfocussed observation categorical, 'If there is a raven, then there is a black raven'). The second tolerates white ravens, if they always happened to be accompanied by a black one. Logically, only the first one requires a cross-referring variable, whereas the second does not in itself require a variable at all.

6. This grossly underestimates the degree to which one relies on testimony, the evidence marshalled by others. We can either regard 'the subject' as an ideal one, representing something like the history of science, or make the point that in order to make to story realistic, we'd need an account of testimony.

7. Actually, a conditional whose antecedent is a pegged observation sentence and whose consequent is a pegged observation sentence is what Quine calls an *observation conditional* (p. 233). Strictly the relation of theory to evidence is that the theoretical sentences imply infinitely many observation conditionals, some of whose antecedents and consequents are actually verified; the relation of theory to empirical content is that theoretical sentences imply the observation categoricals. The categoricals for our purposes can be thought of as generalisations of the observation conditionals. I have suppressed this complication as it does not seem relevant here.

8. One might question this on the grounds that Quine is committed to the thesis of ontological relativity. Quine however thinks that ontological relativity is consistent with this, for since, once again, our view of reality is as always within our conceptual scheme, within our theory, changing one's theory – even in the easy way represented by proxy-functions – is a change of theory all the same. But I'll leave this issue aside (Quine, 1991: 31–36; 2008: 405–406).

9. Frege famously compared the microscope (formal logic, i.e. the 'Begriffsschrift') and the hand (ordinary language) (Frege, 1967 [1879]: 6).

10. It does require that we back off from what Quine says at 1992: 100, where he asserts that if one's theory is A then the claims of B must be 'meaningless'.

11. The closest Quine comes to a serious foray into transcendental metaphysics is when he envisages the use of a Tarkian truth-predicate for the meta-perspectival task of comparing, say, Einsteinian physics with its Newtonean precursor (Quine, 1991: 81). Inter-theoretic equivalence can seriously be conceived semantically: a statement or group of statements is empirically equivalent to another just in case it (semantically) implies just the same observation sentences as the other. Such a comparison is often useful theoretically for the largeness of vantage point it affords. But although the difference in such a case is big one it is not a radical one, as mentioned above.

References

Ben-Menahem, Y. (2006) *Conventionalism: From Poincare to Quine* (Cambridge: Cambridge University Press).

Bergström, L. (1993) 'Underdetermination of Physical Theory'. In R. Gibson (ed.) 1993, *The Cambridge Companion to Quine*. Cambridge: Cambridge University Press, pp. 91–114.

Carnap, Rudolf. (1928) *Der Logische Aufbau der Welt*. Leipzig: Felix Meiner Verlag. Tr. by R. George, 1967, *The Logical Structure of the World. Pseudoproblems in Philosophy*. Berkeley: University of California Press.

Duhem, Pierre. (1954 [1906]) *The Aim and Structure of Physical Theory*, tr. by P. Weiner (Princeton: Princeton University Press).

Frege. G. (1967 [1879]) 'Begriffsschrifft, a formal language, modelled upon that of arithmetic, for pure thought'. In J. van Heijenoort (eds.) 1967, *From Frege to Gödel: A Source Book in Mathematical Logic, 1879–1931*. Cambridge, MA: Harvard University Press .

Gibson, R. (1991) 'More on Quine's Dilemma of Underdetermination', *Dialectica*, 45(1): 59–66.

Howard, D. (1990) 'Einstein and Duhem', *Synthese*, 83: 363–384.

Hylton, P. (1993) 'Quine on Reference and Ontology'. In R. Gibson (ed.) 1993, *The Cambridge Companion to Quine*. Cambridge: Cambridge University Press, pp. 115–150.

Hylton, P. (2007) *Quine* (Oxford: Routledge).

Johnsen, B. (2014) 'Observation'. In G. Harman & E. Lepore (eds.), *A Companion to W.V.O. Quine*. Oxford: Wiley-Blackwell, pp. 333–349.

Kemp, G. (2012) Quine versus Davidson: Truth, Reference and Meaning (Oxford: Oxford University Press).

Moore, A. (2012) *The Evolution of Modern Metaphysics: Making sense of things* (Cambridge: Cambridge University Press).

Neurath, O. (1959 [1934]) 'Protocol Sentences'. In A.J. Ayer (ed.) 1959, *Logical Positivism*. New York: The Free Press, a division of Macmillan Publishing Co., pp. 199–208.

Quine, W.V. (1961 [1953]) *From a Logical Point of View*, second edition (Cambridge, Mass: Harvard University Press).

Quine, W.V. (1969) *Ontological Relatively and Other Essays* (New York: Columbia University Press).

Quine, W.V. (1970) 'Grades of Theoreticity'. In L. Foster & J.W. Swanson (eds.) 1970, *Experience and Theory*. Amherst: University of Massachusetts Press.

Quine, W.V. (1976) *Ways of Paradox*, revised edition (Cambridge, Mass: Harvard University Press).

Quine, W.V. (1981) *Theories and Things* (Cambridge, Mass: Harvard University Press).

Quine, W.V. (1986) *Philosophy of Logic* (Cambridge, Mass: Harvard University Press).

Quine, W.V. (1992) *Pursuit of Truth*, second edition (Cambridge, Mass: Harvard University Press).

Quine, W.V. (1995) *From Stimulus to Science* (Cambridge, Mass: Harvard University Press).

Quine, W.V. (2008a) *Confessions of a Confirmed Extensionalist and Other Essays*. D. Føllesdal & D. Quine (eds.). Cambridge, Mass: Harvard University Press.

Quine, W.V. (2008b) *Quine in Dialogue*. D. Føllesdal & D. Quine (eds.). Cambridge, Mass.: Harvard University Press.

Quine, W.V., & Ullian, J. (1978) *The Web of Belief* (New York: McGraw-Hill Higher Education).

Schlick, M. (1959 [1933]) 'The Foundation of Knowledge'. In A.J. Ayer (ed.) 1959, *Logical Positivism*. New York: The Free Press, a division of Macmillan Publishing Co., pp. 209–227.

Severo, R. (2008) '"Plausible insofar as it is intelligible": Quine on underdetermination', *Synthese*, 161: 141–165.

Sinclair, R. (2014) 'Quine on Evidence'. In G. Harman & E. Lepore (eds.) 2014, *A Companion to W.V.O. Quine*. Oxford: Wiley-Blackwell, pp. 350–372.

van Fraassen, B. (1980) *The Scientific Image* (Oxford: Oxford University Press).

13
Quine, Wittgenstein and 'The Abyss of the Transcendental'

Andrew Lugg

During the early decades of the Twentieth Century many philosophers, W.V. Quine and Ludwig Wittgenstein among them, repudiated what they deemed the pretentions of past philosophy, in particular the assumption that there is knowledge about the world deeper than the deliverances of science and common sense. Attempts to provide information about what really exists, really happens, really matters were judged misguided and metaphysical and epistemological theorizing in all its manifestations decried (compare Carnap, 1996 [1935]: 32). Inquiry that is neither purely scientific nor purely logical was given short shrift, the easy acceptance of philosophical speculation was viewed with suspicion, and much philosophy once supposed legitimate was more or less quietly dropped. In more than a few quarters it became an article of faith that there is nothing deserving the name of philosophical knowledge, only mundane knowledge about the way things happen to be and how such knowledge is acquired. Instead of treating knowledge as falling under three heads – to comprise scientific truths, logical truths and truths lying somewhere between the two – mainstream philosophers took it to fall under two heads – to be either scientific or logical.

Quine and Wittgenstein were as firm as anyone in discounting the idea of a subject midway between logic and science and writing off philosophical speculation of the sort previously trafficked. They worked with a picture of a philosophical Grand Canyon full of questionable notions and disreputable conjectures, it being central to their thinking that philosophical claims that do not survive scientific and logical scrutiny, those about being, knowledge and mind above all, should be given wide berth. Thus in 'Things and their place in theories', the lead article of the last collection of essays that Quine compiled, various important philosophical claims are consigned to 'the abyss of the transcendental' (1981: 23), and in a letter dated January 16, 1916 Wittgenstein says (in connection with the possibility of one sort of philosophical theorizing): '[L]et's cut out the transcendental *twaddle* [*transzendentales* Geschwätz] when the whole thing is as plain as a sock on the jaw' (Engelmann, 1967: 11). While Wittgenstein may have condemned traditional philosophizing more stridently than Quine and was quicker to speak of it as nonsense, Quine was no less in doubt that philosophers of old overstep the mark, that their theories are indefensible, indeed without meaningful content.

In 'Things and their place in theories' Quine refers to four traditional philosophical conceptions as lying in the abyss of transcendental (1981: 22–23). One is 'the reality of the external world – the question of whether or in how far our science measures up to the *Ding an sich*', a second the radically skeptical claim that we know nothing about the external world, a third the idea of a purely 'rational reconstruction of the world from sense data' and a fourth the philosopher's notion of 'a matter of fact'. In addition, elsewhere he targets 'the myth of a museum in which the exhibits are meanings and the words are labels' (1968: 27), deprecates in no uncertain terms the philosopher's use of 'proposition', 'belief', 'thought', 'meaning', and 'experience' (1981: 184–185), writes: 'Epistemology, or the theory of knowledge, blushes for its name' (2008: 322) and asks regarding an attempt to construct 'physical discourse in terms of sense experience, logic and set theory': '[W]hy all this creative reconstruction, all this make-believe?' (1968: 75). He even insists that the old concepts of truth and ontology 'belong to transcendental metaphysics', there being no sense to be made of them in the absence of 'a broader background', the very thing traditional philosophers would abstract from (1968: 68).

And likewise Wittgenstein takes the concepts and theories of traditional philosophy to be beyond the pale. However different his terms of criticism are from Quine's, he comes to much the same conclusions. Nailing his colors to the mast early on, he declares that '[m]ost propositions and questions, that have been written about philosophical matters, are not false but senseless [*unsinnig*]', in fact 'of the same kind as the question whether the Good is more or less identical than the Beautiful' (1990 [1922]: 4.003). And in later work, while somewhat less forthright, he continued to maintain that traditional philosophy is intellectually bankrupt. He deplores what he takes to be the philosopher's penchant for positing '*spiritual* [geistige] activity corresponding to … words' (1953: §36), condemns how 'proposition', 'word', and 'sign' are wielded in philosophy (§105), criticizes philosophical conceptions of sentence and language (§108) and deplores the manner in which 'knowledge', 'being', 'object', 'I', 'proposition', and 'name' are treated in philosoph-ical discussion (§116). Furthermore, like Quine, he regards traditional epistemology as barren. As he sees it, philosophers from the Greeks on fall into the trap of 'talking about knowledge' instead of 'talking about particular instances of knowing' (2005: 54).

Quine and Wittgenstein's indictment of metaphysical and epistemo-logical speculation could scarcely be more sweeping and uncompro-mising. They hold that philosophical notions and hypotheses cannot be justified by intuition, rational insight, or sustained reflection, not even by showing them to be 'grounded' in science or language. Still, neither philosopher dismisses past philosophy entirely, never mind opts for the comfortable option of rejecting the discipline root and branch. They allow – in Wittgenstein's words – that the problems of philosophy have 'the character of *depth* [*Charakter der* Tiefe]' (1953: §111) and attempt to do justice to their depth, each in his own fashion and for his own special purposes. Rather than close their eyes to the tradition, they take them-selves to be 'destroying...houses of cards and...clearing up the ground of language on which they stand' (§118). Both think 'the impressiveness [of philosophy] retreats to [the] illusions, to the problems', to philoso-phers' worries rather than to the theories they advance (§110). It is this, I believe, that defines their efforts and constitutes their main claim to fame. Their importance lies in how they handle the verdict that philos-ophy cannot be done the way it used to be done, not in their turning their backs on the subject as traditionally practiced.

Where Quine and Wittgenstein primarily differ is over how they respond to and provide for their shared view of past philosophy. Quine treats philosophy, some of it at least, as science rather than as standing above, below or alongside science, while Wittgenstein bends his energies to exposing the misunderstandings he sees philosophy as plagued by. On the question of how knowledge should be thought of, for instance, Quine responds by considering how 'know' can be bent to technical use, Wittgenstein by considering how it and its cognates are used. The differ-ence is that Quine takes up the question with the aim of contributing to the science of science whereas Wittgenstein takes it up with the aim of untying knots in which the philosophical-minded are all-too-liable to find themselves entangled. Each circumvents the difficulties that he takes to beset traditional philosophy by eschewing philosophical speculation and scouting an alternative to the subject as traditionally practiced, all the while mindful of the concerns that have motivated philosophers' inquiries down the centuries. (There is a little more on this theme in the Appendix at the end of this paper.)

It is basic to Quine's way of thinking that certain philosophical conceptions do not have to be left in the abyss of the transcendental but can be usefully extracted from it. Thus in 'Things and their place in theories' he recasts 'the question whether or in how far our science matches up to the *Ding an sich*' as a question about the degree to which scientific theory is confirmed by observation. He observes that radical skepticism stems from confusion but is 'not of itself incoherent', science being 'vulnerable to illusion on its own showing, what with seemingly bent sticks in water and the like'. He points out that 'the project of a rational reconstruction of the world from sense data' regarded as 'similarly naturalistic' fails, there being no possibility of constructing 'a language adequate to natural science' from 'a realm of posited entities intimately related to the stimulation of sensory surfaces'. And he states that the notion of a matter of fact is 'not transcendental or yet epistemo-logical, not even a question of evidence' but 'ontological, a question of reality,... to be taken naturalistically within our scientific theory of the world' (1981: 22–23). His strategy is to construe philosophers' notions and theories as far as possible so that they pass scientific muster. As he observes when discussing 'matters of fact', he is 'at pains to rescue [such notions and theories] from the abyss of the transcendental'.

For the most part Quine is out to clarify the transition 'from stimulus to science' (1995) with an eye to explaining how we acquire our theory of the world given the meager input provided by the senses. He reformu-lates the philosopher's 'transcendental' question as 'immanent' (for him 'transcendental' and 'transcendent' are synonymous nearly enough) and he sets about showing how '[i]n fused phrases of Kant and Russell,... our knowledge of the external world is possible' (1992 [1990]: 18). Rather than attempt to reconstruct our knowledge from scratch (and without appealing to scientific fact or theory), he attempts, as he says in another late work, to provide a 'rational reconstruction of the individual's and/ or the race's actual acquisition of a responsible theory of the external world' (1995: 16), one that is empirically true or false (compare 1975: 70). His aim is to demonstrate that our knowledge of the external world can be explained without invoking the philosopher's stock notions of proposition, meaning and translation (except insofar as they are scien-tifically testable).

There is also a sense in which Wittgenstein aims to rescue the concepts and claims of past philosophy. He does not recommend treating them as scientific but would have them be construed, more prosaically, as they are construed day in, day out in ordinary life. Thus in the *Investigations* he counsels philosophers to retrieve the normal uses of 'knowledge', 'being', 'object', 'I', 'proposition', and 'name' (1953: §116). 'What we do', he writes (referring to philosophers who proceed as he does), is 'bring words back from their metaphysical to their everyday use'. Instead of going along with how philosophers use words, he juxtaposes their use with how they are used in their 'original home', his hope being that this will reveal philosophers' claims to be logically indefensible. He does not ridicule past philosophers' efforts – indeed he sees some as documenting 'a tendency of the human mind which [he] cannot help respecting deeply' (1993: 44) – but takes it upon himself to show that philosophy understood positively is speculative in the extreme, actually little better than mystery mongering, idle chatter that obscures rather than sheds light on how things are.

Effort of the kind Wittgenstein champions, unlike effort of the kind Quine champions, is not intended to add to our store of positive knowledge about the world, language or the mind. (In Wittgenstein's hands the study of grammar is a descriptive, not an explanatory, endeavor.) Whereas Quine seeks to contribute to our understanding of how we come by our language and our theory of the world, he seeks – by the not-so-simple expedient of reminding us of what we already could, should or do know – to convince us that philosophical talk is empty. He is uninterested in the scientific study of the acquisition of scientific theory and concentrates on philosophical questions that 'produce in us mental cramp' (1958: 1). His hunch is that philosophical disputes, especially metaphysical and epistemological disputes, arise because 'philosophers constantly see the method of science before their eyes', more specifically 'crav[e] ... generality' and manifest a 'contemptuous attitude towards the particular case' (1958: 18; also compare 1979: 68 and 1993: 44). Persuaded that philosophy is riddled with muddles, he endeavors to isolate the 'superstitions' engendered by 'grammatical illusions' (1953: §109) and demonstrate how 'to pass from a piece of disguised nonsense to something that is patent nonsense' (§464). In his view enlightenment in philosophy comes from confronting what is already known and he devotes himself to 'dispers[ing] the fog', the 'haze which makes clear vision impossible' (§5).

So far my argument has been that Quine and Wittgenstein proceed from the same starting point in conspicuously different ways. Quine responds to the failure of old-school philosophy by offering and exploring a scientific replacement for a traditional philosophical problem while Wittgenstein responds by exploring the ins-and-outs of what he takes to be the thinking behind philosophical problems with the object of showing what he takes to languish in the abyss of the transcendental belongs there (and disabusing us of what seems to be apparent inexorability of a range of philosophical views). In other words, I am suggesting that Quine's approach is two-fold, Wittgenstein's one-fold. They both repudiate the concepts and theories of past philosophy differently (Quine does not duck the task of showing this), but whereas Wittgenstein confines himself to weaning us from traditional philosophy, Quine goes on to develop a scientific alternative. Viewed from this perspective, the main difference between them is that Quine's treatment of philosophical speculation is less detailed than Wittgenstein's, and there is nothing in Wittgenstein's writings comparable to Quine's account of the transition from stimulus to science. This much should be uncontroversial, it being hard to see how else what they say and how they describe what they are doing can be satisfactorily accounted for.

More controversial is how exactly Quine's and Wittgenstein's investigations ought to be understood. It is not unreasonable to view their writings, as they are typically viewed, as standing in the sharpest of contrasts, as different as chalk and cheese. There is a world of difference between believing '[p]hilosophy is in large part concerned with the theoretical, non-genetic underpinnings of scientific theory' (Quine, 1976: 151) and believing '[p]hilosophy simply puts everything before us, and neither explains nor deduces anything' (Wittgenstein, 1953: §126). This by itself, however, falls well short of proving that Quine's and Wittgenstein's projects are mutually exclusive. Seeing that they proceed differently is not tantamount to seeing them as fundamentally opposed, and it is far from a foregone conclusion that one of their approaches is objectively superior. Chalk and cheese are not in competition, and it has to be shown that Quine and Wittgenstein promote conflicting alternatives, that just one of their approaches at most is tenable. To steal a favorite phrase of Quine's, it remains to be established that there is a 'fact of the matter' who is right and who wrong. The key question – one rarely entertained, let alone discussed – is what exactly the difference between Quine and Wittgenstein comes to.

My answer to this question is that when Quine and Wittgenstein are read, as I am claiming they have to be read, as agreeing that philosophizing in the old manner is no longer possible, there is no escaping the conclusion that they proceed in compatible ways. Their projects are not in conflict but incommensurable in the sense that there is no saying one is philosophically superior to the other. Quine is primarily engaged in scientific theory construction, Wittgenstein in criticizing traditional philosophy, i.e. Quine can be regarded as salvaging the rubble to build something new, Wittgenstein as clearing away the rubble to remove an eyesore. Put yet another way I am suggesting that Quine and Wittgenstein differ in interest and attitude, not doctrine and belief. They are best understood as following in different ways what Wittgenstein refers to in the *Tractatus* as 'the right method in philosophy', 'the only strictly correct method' (1990 [1922]: 6.53). Quine tacitly accepts the injunction '[t]o say nothing except... the propositions of natural science' while Wittgenstein would demonstrate to whomever wishes 'to say something metaphysical,... that he has given no meaning to certain signs in his propositions'. (The complaint that Quine's theorizing has 'nothing to do with philosophy' and Wittgenstein's reflections are philosophically 'unsatisfying' is discussed in the Appendix.)

Neither Quine nor Wittgenstein explicitly allows for the kind of philosophy the other advocates, but there is reason to think that they would not repudiate it (at least not as I have been describing their projects). Though Quine had little good to say about ordinary language philosophy and took Wittgenstein to be one of its leading lights (1968: 82), he did not hesitate to appeal to common usage when it served his purposes and was on occasion as swift as Wittgenstein to suggest that philosophers go astray because they misuse language (1976: 229). (It is even arguable that Quine was, if anything, more attentive to ordinary usage than Wittgenstein.) And for his part Wittgenstein was not opposed to properly-conducted scientific investigation (as distinct from philosophical speculation masquerading as scientific fact). Had he had the opportunity to review Quine's remarks, he may well have objected to some, even many, of them (and protested that positive and negative considerations are mixed together). But he would not have criticized Quine's favoring scientific theorizing over philosophical speculation. It is only when Quine's and Wittgenstein's discussions are considered apart from their joint criticism of past philosophy – and overmuch consideration is paid to their rhetoric – that they seem engaged in opposed endeavors.

There are also fairly clear hints in Quine's and Wittgenstein's writings that they recognize the possibility of different ways of negotiating the abyss of the transcendental. However strongly they endorsed their own ways of philosophizing, they do not exclude other ways of proceeding. To the contrary, they occasionally accept, if only obliquely, that the difference between their approaches is rooted in philosophical preference rather than philosophical commitment. Quine recognizes that 'Wittgenstein's characteristic style, in his later period, consisted in avoiding semantic ascent [i.e. "talk of theory" instead of "talk within theory"] by sticking to the examples' (1960: 274, footnote 4). And Wittgenstein reportedly allowed (after saying how ill-disposed he was to 'idol worship, the idol being Science and the Scientist') that he was 'in a sense making propaganda for one style of the thinking as opposed to another' (1966: 27–28). Such remarks do not prove that Quine and Wittgenstein recognize alternative responses to the demise of the tradition. They do, however, indicate that each countenances the possibility of more than one response. It cannot be entirely by chance that both describe a way of doing philosophy as a 'style'.

What intrigued Quine did not intrigue Wittgenstein and what intrigued Wittgenstein did not intrigue Quine. Beyond the fact that their philosophical backgrounds did not fully overlap and the philosophical settings in which they worked were poles apart, they were temperamentally very different sorts of thinker. Quine would never have said as Wittgenstein did: 'Scientific questions may interest me, but they never really grip me [*nie wirklich fesseln*]', and Wittgenstein was much more concerned than Quine with '*conceptual* & *aesthetic* questions' (Wittgenstein, 1998: 91). Unlike Wittgenstein, who confesses to being uninterested in 'whether scientific questions are solved', Quine was exercised by scientific questions, those associated with his own epistemological project in first place. He was not riveted, as was Wittgenstein, by '[t]he decisive movement in the [philosophical] conjuring trick' (1953: §308), and Wittgenstein was not riveted, as was Quine, by the technical problem of providing a scientifically acceptable account of how we come to have 'a responsible theory of the external world'. Each was chiefly motivated by his own predilections, an interest in science in Quine's case, an interest in intellectual sleight-of-hand in Wittgenstein's, a difference that is hardly subject to objective arbitration.

In my view, then, there is no conflict between Quine's attempt to contribute to knowledge by recasting problems of traditional epistemology as scientific problems and Wittgenstein's attempt to liberate us from misbegotten speculation and wishful thinking by uncovering and sorting out philosophical muddles. Explaining how individuals singularly and collectively acquire language and their theory of the world is different from unmasking irresponsible thinking about the world and our place in it, and however different the two projects, they are not incompatible. Construction (even construction preceded by criticism) is not criticism, and it is a mistake to think philosophers have to choose between them if they are to embrace either. Providing a satisfying – naturalistic – account of the language and scope of science (Quine, 1976: 228ff) does not exclude making 'philosophical problems...*completely* disappear' (Wittgenstein, 1953: §133). Both tasks can be intelligibly pursued and not only by different philosophers, Quine's discussion of radical translation being perhaps the plainest example of philosophical criticism and scientific theorizing being carried out in tandem (1960: chapter 2).

But is it true that Quine's and Wittgenstein's goals and tactics are compatible and their investigations should be regarded as running in parallel? It is tempting to argue against this that what they say sometimes coincides, sometimes diverges, and hence only one of their projects can be right. There are many passages in their writings in which they appear to espouse the same views, even express them in the same words, and many passages in which they espouse what appear to be opposed views, even by the look of it to be contradicting one another. This makes it seem as though they have the same philosophical goals, even that they are engaged in the same enterprise as Plato, Descartes, and Hume. Whether this is so, however, is to say the least moot. To show that Quine and Wittgenstein are in the same line of business, it is not enough to point out that their discussions intersect and they sometime seem to be agreeing, sometimes disagreeing. The stronger conclusion follows only if the appearance of agreement and disagreement does not melt away when their differing ways of negotiating the abyss of the transcendental are factored in (and they do not have the same ends in sight when their background assumptions are taken into account).

Consider for instance how Quine and Wittgenstein treat the all-important topic of meaning, a topic on which they are regularly thought to concur more than superficially. There can be no denying that both reject the idea of meanings as entities and take meaning to be connected with use and not only because Quine applauds Wittgenstein's view that 'the meaning of a word is to be sought in its use' (1981: 46). But what they have in mind is very different, their appearing to say the very same thing notwithstanding. It is insufficiently well-appreciated, perhaps not even by Quine himself, that there is an enormous gulf between them regarding the concepts of meaning and use (compare Hacker, 1996: 207–211). It is not just that Quine analyzes names away (1960: 181ff) while Wittgenstein holds that 'the *meaning* of a name is sometimes explained by pointing to its *bearer*' (1953: §43). They differ more deeply on 'use'. For Quine it is a theoretical concept, one belonging to his scientific investigation of language, for Wittgenstein a critical tool, invaluable for exposing the shortcomings of traditional philosophical speculation. It is unimaginable that Wittgenstein would say (as Quine does): 'This is where the empirical semanticist looks: to verbal behaviour' and allow that 'we can take the behaviour, the use, and let the meaning go' (1981: 46).

Much the same can be said about texts in which Quine and Wittgenstein launch similar attacks on traditional philosophical speculation. While both are critical of past philosophy, they have different goals and only travel along a single track a short distance. In a striking passage, for instance, Quine treats a philosophical debate over the existence of miles in a manner reminiscent of nothing so much as one of the debates Wittgenstein details in the *Investigations* (1960: 272). It was Quine but it could have been Wittgenstein who envisions a proponent of miles saying: 'Of course there are miles. Wherever you have 1760 yards, you have a mile', a skeptic answering: 'But there are no yards either. Only bodies of various lengths', and the proponent of miles responding in turn: 'Are the earth and moon separated by bodies of various lengths?' This, however, is where the similarity ends. Whereas Wittgenstein would follow such a debate through many iterations, Quine simply says: 'The continuation is lost in the jumble of invective and question-begging. When on the other hand we ascend to "mile" and ask which of its contexts are useful and for what purposes, we can get on; we are no longer caught in the toils of our opposed uses'. His object is to clear the way for his own positive views, Wittgenstein's to show that philosophical debates do not go anywhere, that they are bizarre, not just tiresome hurdles to be overcome.

Similar observations are equally in order for cases in which Quine and Wittgenstein seem to be offering competing answers to the same question and hence once more seem to have the same philosophical objectives. Appearances are again deceptive, and what seem to be substantial divergences between them turn out to be a result of their having different concerns rather than different convictions. Thus regarding the question of how children acquire words, they seem for all the world to have opposed views of ostensive teaching and differ on whether 'sepia' can be taught straightforwardly by pointing to a reddish-brown sample. When Quine says: 'The colour word "sepia", to take one of [Wittgenstein's] examples, can certainly be learnt by the ordinary process of conditioning, or induction' (1968: 31), he is naturally read as challenging Wittgenstein's view that 'an ostensive definition can be variously interpreted in every case' (1953: §28). In fact, however, there is no conflict. Quine cites the example in the course of his on-going scientific inquiries, Wittgenstein in the course of criticizing traditional philosophical thinking. Whereas Quine is concerned with the way in which language is acquired and 'the ordinary process of conditioning', Wittgenstein is concerned with the philosopher's conception of how 'the ostensive definition "That is called 'sepia'" [helps a person] to understand the word' (1953: §30).

Two further cases of Quine and Wittgenstein appearing to differ may serve to reinforce the point. One is Quine's suggestion that it is odd that Wittgenstein and legions of fellow thinkers should 'stoutly maintain that "true" [and] "exist" [are] ambiguous' rather than, as he would have it, univocal (1960: 131). This seems to be directed at those on Wittgenstein's side of the fence but is better read as expressing a different point. Quine is noting that just one sense of each word is required for a (regimented) statement of scientific knowledge, not objecting to Wittgenstein and like-minded thinkers' view that the two words are deployed in more than one way in everyday speech. And it is just as wrong to regard Quine and Wittgenstein as having opposed views regarding description and explanation. Quine would not say: 'We must do away with all *explanation*, and description alone must take its place' (Wittgenstein, 1953: §109; also §128), but he is as unbending in his repudiation of philosophical explanation as Wittgenstein. It is one thing to think, as Quine does, that explanation is the acme of scientific investigation and description a crucial scientific supplement, quite another to insist, as Wittgenstein does, that explanation is the bane of philosophy and description its saving grace.

These considerations could be amplified and supplemented to cover other seeming similarities and differences, not least Quine's remarks about ontological relativity (1992 [1990]: 50–52) and Wittgenstein's claim that '[w]e want to establish an order in our use of language' (1953: §132). By now, however, the point should be clear. There is no concluding from the fact that on occasion Quine and Wittgenstein appear to espouse the same views and on occasion appear to espouse different views that they are in substantial agreement or disagreement. Their remarks cannot be divorced from their aims and background assumptions but have to be understood in the context of their respective ways of responding to what they take to be the failings of traditional philosophy (and their different philosophical interests). Given how they negotiate the abyss of the transcendental, they were pretty much bound to appear as if they both converge and diverge in their thinking. What they mean is a function of their special perspectives just as the meaning of a mark on a line is a function of the background system of representation (with a single mark variously indicating, say, a spatial location or a spectral color and two marks indicating spatial separation or a mixture of colors).

Still, even granting Quine and Wittgenstein are engaged in different projects, is there not a fact of the matter that upsets the balance and shows Quine's thinking to be objectively superior to Wittgenstein's or Wittgenstein's objectively superior to Quine's? One natural thought here is that Quine has the edge since he acknowledges that science can provide insights useful to philosophers, something Wittgenstein refuses to acknowledge. There is something to this. Wittgenstein has little good to say about science and insists there is no place for scientific investigation in philosophy (1966: 27). But irrespective of how skeptical Wittgenstein was about the importance of science and scientific results for philosophy, Quine goes too far when he characterizes him as encouraging 'steadfast laymanship' (1960: 261). From beginning to end Wittgenstein held with minor variations that '[t]he totality of true propositions is the total natural science' (1990 [1922]: 4.11) and would have agreed that scientific analysis may occasionally be more effective in dissolving a philosophical problem than conceptual analysis. What he deplored was the tendency of philosophers to conflate philosophy with science and the comparable tendency of scientists to shore up their conjectures with philosophical speculation (1993: 274–275).

Quine might also be thought to come out ahead because his conception of philosophical explication is less restrictive than Wittgenstein's and he believed philosophers should treat the problems that concern them in a correspondingly more realistic fashion. In this regard his (scientific) explication of the notion of an ordered pair, an explication he considers a 'philosophical paradigm', may be entered in evidence (1960: 257). But leaving aside the question of how paradigmatic the explication is, why suppose Wittgenstein would have found it problematic? The explication may pose a problem for philosophers who focus on 'the subtle irregularities of ordinary language' (259) but Wittgenstein would not have been fazed, it being no part of his brief that philosophers should confine themselves to subtleties of ordinary language. He was not hostile to scientific analysis and would, I imagine, have welcomed an analysis of 'ordered pair' that showed it to be metaphysically inconsequential. Whatever Quine may have assumed, Wittgenstein did not regard ordinary usage as a be-all and end-all but aimed to find a way beyond philosophical blind alleys of every description. He was not against technical explication and should not be upbraided for 'failing to appreciate that it is precisely by showing how to circumvent the problematic parts of ordinary use that we show the problems to be purely verbal' (Quine, 1960: 261).

Yet another way Quine's approach is only apparently better than Wittgenstein's has to do with the very possibility of conceptual analysis. An account of our acquisition of language (and scientific theory) as foreseen by Quine would not obviate the possibility of grammatical truth of the sort Wittgenstein thinks philosophers should strive to provide. Whether or not '[d]eeper insight into the nature of scientific inference and explanation may someday be gained in neurophysiology' (Quine, 2008: 400), an investigation of language still makes sense. Grammar remains grammar and science remains science whatever the prospects for a scientific theory of the transition from stimulus to science. One only has to recall Quine's analysis of the notion of an ordered pair and his praise for Einstein's discussion of simultaneity (1960: 272) to appreciate that grammatical inquiry is integral, not antithetical, to scientific epistemology. It is no accident that Quine keeps a close eye on the niceties of normal grammar, recognizes a conception of analyticity of the sort Wittgenstein favors (1974: 79–80) and reckons that the 'essentials [of the relation of evidential support] can be schematised by means of little more than logical analysis' (1992 [1990]: 2).

By the same token Wittgenstein cannot be regarded as faring better than Quine. It is mistakenly, if commonly, thought he is on firmer ground since he denigrates scientism and would reject out of hand the suggestion that 'philosophy of science is philosophy enough' (Quine, 1976: 151). While less friendly to the arts than Wittgenstein, Quine was not hostile to them. He did not regard scientific inquiry as the alpha and omega of intellectual life, only took it to be exclusively pertinent given his epistemological concerns. He emphasizes scientific prediction since he 'see[s] it as defining a particular language game, in Wittgenstein's phrase: the game of science, in contrast to other good language games such as fiction and poetry' (1992 [1990]: 20). For him, as he observes in response to the charge that he glorifies science, his 'scientism' boils down to the view that philosophy should be pursued 'as part of one's system of the world, continuous with the rest of science' (1998: 430), and the reason he thinks philosophy of science is all that philosophers require is that no more is needed when it comes to 'certain problems of ontology, say, or modality, or causality, or contrary-to-fact conditionals' (1976: 151).

Again it is no argument for Wittgenstein's approach that it is senseless to hanker after a theory of our possession of language and scientific theory. Wittgenstein is not alone in holding that '[i]n psychology there is what is problematic and there are experiments which are regarded as methods of solving problems' and believing that the latter 'quite by-pass the thing that is worrying us' (Wittgenstein, 1980, §1039). But there is also an argument to be made for the view that the problem that worries Quine admits of a scientific solution, an argument that Wittgenstein would readily accept. On the face of it the project of explaining the transition from stimulus to science scientifically is perfectly coherent and no amount of philosophical scrutiny of Quine's words (as opposed to further scientific investigation) can show it to be a pipedream. For Quine explaining how we come to have language and our view of the world is in principle no more problematic than explaining perambulation and how we digest our food. And who is to say he is wrong? If he were an 'ordinary science philosopher' who supplemented traditional philosophy with scientific theory, he could perhaps be criticized, even decisively. But he is not such a philosopher and he cannot be so criticized.

And finally Wittgenstein is not justifiably preferred because he keeps his counsel whereas Quine offers a theory of how we acquire language and our theory of the world of questionable scientific merit. It is wrong to read Quine as palming off a philosophical doctrine as a scientific theory and defending a newfangled version of empiricism, one not essentially different from the old-fashioned version. Empiricism for him is a scientific doctrine, and his materialism summarizes an important scientific result. No doubt he invites criticism when he baldly states (in an important early statement of his thinking about language and scope of science) that he is 'a physical object sitting in a physical world' (1976: 228). But while the notions of 'physical object' and 'the physical world' are reasonably regarded as prime examples of the sort of notion to found in the abyss of the transcendental, Quine himself is not peddling a philosophical doctrine. While he does not belabor the point, he clearly means to be reporting a scientific fact. As he understands the offending notions, they are genuinely scientific (and subject to revision, even rejection).

The purpose of the discussion up to now has been to turn back common misunderstandings of what Quine and Wittgenstein are about and how they negotiate the abyss of the transcendental. My central claim has been that they should be read as pioneering complementary approaches to traditional philosophy, not – as they are standardly read – as engaged in what usually passes for philosophy. I have not attempted to detail what they argue but have confined myself rather to bringing out what I see as their main contributions to philosophy. As I interpret them, Quine is a major philosopher because he never wavers in his determination to replace rather than supplement traditional philosophy and Wittgenstein a major philosopher because he never wavers in his determination to dissolve philosophical problems. Only when they are viewed as responding to the demise of traditional philosophy and as discounting the possibility of solving the old problems by citing facts – scientific facts in Quine's case, grammatical ones in Wittgenstein's case – is the full power of what they are urging apparent and easy objections to it turned aside.

When Quine and Wittgenstein are viewed as shunning what they take to be outmoded philosophical speculation, much more of what they say falls into place and is not subject to trivial complaint. It is no surprise that a philosopher engaged in a scientific-cum-epistemological endeavor should treat matters from a third-person scientific perspective and demand that notions be scientifically acceptable, and no surprise that a philosopher engaged in a critical-cum-dialectical enterprise should challenge the assumption that words have defining characteristics and question the possibility of a language logically inaccessible to all but the speaker. Quine's characterization of objects as 'posits' no longer seems bizarre (1960: 22), his point that 'analytic' belongs to a circle of terms no longer seems ham-fisted (1980 [1953]: 32) and his analysis of the notion of a natural kind in terms of the idea of 'an innate standard of similarity' no longer seems hopelessly inept (1968: 123). Nor can Wittgenstein's criticism of private ostension be quickly dismissed as begging the question (1953: §§28ff), his discussion of what is common to all games, all numbers and the like be condemned as shaky in the extreme (§§66ff) or his suggestion that the meanings of words are fixed by the speaker's community be treated as obviously untenable (§§199ff).

And it is no mystery either, when Quine and Wittgenstein are read in the way I am claiming they have to be read, why they paid no heed to objections of the kind widely thought to show they do not have a leg to stand on. Neither philosopher was oblivious to commonly-stated criticisms of his views and never recanted, proceeding instead as though they did not warrant his attention. They are philosophers of the first rank and it is, I would say, no little argument in favor of the interpretation of their thinking that I am defending that they do not make elementary mistakes of reasoning or fail to address important philosophical questions. Quine does not botch the topics of necessary truth, ontology and empiricism and behaviorism (Dilman, 1984: viii), and Wittgenstein does not end up espousing forms of anti-realism and relativism that nobody should espouse (Grayling, 1988: 100–109). Quine had reason to believe – given his replacement of traditional philosophy by science – that the notion of necessity can be captured in non-modal terms, that ontology can be usefully studied and that empiricism and behaviorism are scientifically sound doctrines. And Wittgenstein had reason to believe – given his critical concerns – that linguistic practice was heterogeneous and what is meant by a word depends on the language-game in which it is woven. Neither philosopher purports to contribute positively to philosophy as usually practiced for the simple reason that neither intends to speculate philosophically.

Nothing I have said is meant to suggest that Quine and Wittgenstein never put a foot wrong and what they say in the course of pursuing their respective approaches is immune to criticism. My contention is the approaches they pursue make good sense and common objections to their thinking fall short. Both philosophers would have allowed that there is work to be done. It is not the view of either that he has everything sorted out, and it is a mistake to dismiss his approach on the grounds that it fails to handle satisfactorily all the problems he tackles. Indeed both Quine and Wittgenstein remained dissatisfied on many details. For instance Quine does not purport to have a fully adequate account of the transition from stimulus to science, a proper account of the nature of observation sentences and the degree to which such sentences are theory-laden being something he was struggling with to the end (see 1992 [1990], §§2–3 and §43, and 2008: 476–477). And in his last writings Wittgenstein was still striving 'to bring the concepts into some kind of order' (1977, II.12), the concepts of impossible colors such as reddish green, pure brown, transparent white and luminous grey forever eluding his grasp (III.131–295). What deserves special consideration, I am contending, is not the specific results of Quine's and Wittgenstein's investigations but the nature of their investigations, how they went about treating the notions and theories they took to languish in the abyss of the transcendental.

I have been intimating and believe there is life in both Quine's and Wittgenstein's philosophical approaches. But I would be the first to concede that their ways of proceeding are not in much favor nowadays and anyone who tried to follow faithfully in their footsteps would have an uphill task. Even philosophers who despair of past philosophy regard Quine's 'scientism' as a step too far and take Wittgenstein's 'critical stance' to be overly negative. Rather than assimilate philosophy to science or treat it as a suitable case for treatment, they hold out for a more positive role for the discipline, one more in line with what past philosophers envisaged. The early pragmatists' view of meaning, truth and value as a matter of practical consequences and the logical positivists' view of philosophy as the logical analysis of science have both been subject to a welter of criticism, and a different way of preserving the manifest advantages of Quine's and Wittgenstein's approaches while avoiding their perceived disadvantages is now sought. Best, evidently, would be a way of straddling the fence that does not, as intermediate views usually do, collapse into one or other of the views it is supposed to supersede.

Though rarely presented as responding to Quine's and Wittgenstein's grim view of past philosophy, any number of recent philosophical efforts can be regarded as aiming to go beyond and improve on their philosophical thinking. Thus, to mention a few examples, it has been suggested that the discipline can be revitalized by mapping our everyday conceptual scheme, by stitching together behavioral and social facts to produce a philosophically acceptable explanation of our day-to-day practice, and by saying something philosophically useful in the form of autobiography or commentary on the passing scene. Nobody can say what the future will bring, and it is by no means obviously that there is no steering between the Scylla of the Quine's scientific maneuvers and the Charybdis of the Wittgenstein's critical reflections. But neither is it obvious that there is philosophical knowledge to be extracted from common-sense thinking, something philosophically significant to be gleaned from psychology and sociology, or philosophical insight into how things are to be had by examining our own special concerns or those of any chosen social group. In the end Quine's and Wittgenstein's ways of negotiating the abyss of the transcendental may be as good as it gets.

I began by noting that in the early decades of the Twentieth Century many philosophers denounced what they judged to be the pretensions of past philosophy, and I would be remiss if I did not mention in concluding that in recent years the pendulum has swung back and very few philosophers see a substantial hiatus between past and present philosophical practice. Today philosophy is almost never regarded as being in the predicament Quine and Wittgenstein take it to be in, and it is seldom argued that fundamental rethinking is required to put it back on its feet. Instead of repudiating philosophical speculation, philosophers devote themselves to contributing, as they see it, to philosophical knowledge by advancing philosophical theories. If the conception of past philosophy as 'twaddle' that belongs in the abyss of the transcendental is considered at all, the assumption seems to be, as Bertrand Russell once put it, that '[t]o attempt the impossible is, no doubt, contrary to reason; but to attempt the possible which *looks* impossible is the summit of wisdom' (1951: 458). Scientific ideas have been retrieved, so why not philosophical ideas? But then again there is the fact that, while the notions of a void and real chance in nature have been retrieved, the notions of phlogiston and absolute simultaneity seem gone forever. The return of philosophical theorizing may be all to the good but it may also be yet another example of the triumph of hope over experience.

Appendix: Are Quine and Wittgenstein still doing philosophy?

Quine and Wittgenstein are regularly portrayed as turning their backs on philosophy itself, not just on philosophy as traditionally pursued. It is argued, especially by philosophers in Wittgenstein's camp, that philosophy naturalized is different from the genuine article (compare Wittgenstein 1923, 6.53: '[N]atural science...has nothing to do with philosophy'). And it is argued, especially by philosophers in Quine's camp, that taking Wittgenstein at his word means accepting that there is no such thing as philosophy, only nonsense (6.53 again: '[T[o demonstrate to [someone who wished to say something metaphysical] that he had given no meaning to certain signs in his propositions...would be unsatisfying...– he would not have the feeling we are teaching him philosophy'). These complaints are not outlandish if only because Quine and Wittgenstein often seem to suggest that philosophy itself should go. But they can also be regarded – as Quine and Wittgenstein themselves regarded them – more generously, as aiming each in his own way to set the subject on a more productive path. Neither holds that philosophy has come to an end, only insists that it should be pursued in a different fashion, one properly thought of as a successor to the subject as traditionally understood.

Quine was aware that many readers took him to be giving up on philosophy but considered the charge to be without merit. He saw himself as advocating a shift in what philosophers can responsibly investigate, not as peremptorily rejecting philosophy. What he was doing, he insists, is not entirely disconnected from what philosophers have been doing all along but connected with it. Thus in the case of our knowledge of the external world, he not only writes: 'Our liberated epistemologist ends up as an empirical psychologist, scientifically investigating man's acquisition of science' and agrees that liberated epistemology is '[a] far cry...from old epistemology', he also avers that it is 'no gratuitous change of subject matter', it being 'an enlightened persistence...in the original epistemological project' (1974: 3). This is a difficult but not an implausible view, and it is worth pondering why it is wrong to protest 'the normative element, so characteristic of epistemology, goes by the board' (Quine, 1992 [1990]: 19). The best response, surely, is to consider how naturalistic philosophy may be, as Quine would have it, continuous with the philosophy of Descartes, Leibniz and other so-called natural philosophers concerned with the nature and limits of knowledge (1981: 190–191).

As for Wittgenstein, his conception of philosophy is just as mistakenly dismissed as unrelated to the traditional conception. Like Quine, he takes his brand of philosophy to be similar to, as well as different from, traditional philosophy. He recognizes that he is shifting the subject but sees a clear link between what he does and what Plato, Aristotle, Descartes and other philosophical luminaries of the past were doing. Thus in a lecture in 1930/1931 he stresses that he is engaged in philosophy despite 'doing a "new subject"', one that is 'not merely a stage in a "continuous development"' but 'a "kink" in the development of human thought' (1993: 113). As he reportedly stated when questioned whether philosophy as he does it 'should be called "philosophy"', the vital fact is that – while what he does is 'certainly very different from what, e.g. Plato and Berkeley had done' – 'people might feel that it "takes the place" of what they had done'. Making sense of this is tricky but not, I am inclined to think, impossible. To hold, as Wittgenstein does, that the shift in philosophical approach he recommends is as big as the shift that occurred when 'Galileo and his contemporaries invented dynamics' and 'when "chemistry developed out of alchemy"' is not to exclude the possibility that his problems and terms of criticism are relevantly similar to Plato's and Berkeley's.

When considering Quine's and Wittgenstein's response to what they judge to be the demise of traditional philosophy, it should not be forgotten that neither of them thinks of philosophy as a single fixed kind of enterprise. Quine regarded Wittgenstein as a philosopher (and not only because he mistakenly viewed him as an ordinary language philosopher), while Wittgenstein, a philosopher unusually unfussy about what does and does not count as philosophy, would have accepted that Quine is doing something reasonably called philosophy (and not only because he would in all probability have judged his 'science' to be philosophy in disguise). Quine applauds what he refers to as 'a casual attitude towards the demarcation of disciplines' and protests that '[n]ames of disciplines should be seen only as technical aids in the organization of curricula and libraries' (1981: 88; also 190), and Wittgenstein regards the multitude of endeavours that go by name of 'philosophy' in much the same way as he regards games and numbers, namely as linked by similarities and differences rather than as sharing as a single defining characteristic (1953: §§66ff). For both philosophers it makes no difference – as Wittgenstein puts it in another context – how something is called 'so long as it does not prevent you from seeing the facts' (1953: §79). What matters is how each of them negotiates the abyss of the transcendental, not whether their thinking is accurately characterized as satisfying this or that predetermined conception of philosophy.

Acknowledgements

The argument of this paper owes much to conversations I had with Burton Dreben during the last years of his life (also compare his under-appreciated and widely misunderstood 1996). In addition I should like to thank Puqun Li and Warren Ingber for going through the penultimate draft and Paul Forster for helpful comments on earlier drafts and continuing useful discussion.

References

Carnap, R. (1996 [1935]) *Philosophy and Logical Syntax* (Bristol: Thoemmes Press).

Dreben, B. (1996) 'Quine and Wittgenstein: The Odd Couple'. In R.L. Arrington & H-J. Glock (eds.) 1996, *Wittgenstein and Quine*. London: Routledge .

Dilman, I. (1984) *Quine on Ontology, Necessity, and Experience*. (Albany: State University of New York Press.)

Engelmann, P. (1967) *Letters from Ludwig Wittgenstein* (Oxford: Blackwell).

Hacker, P.M.S. (1996) *Wittgenstein's Place in Twentieth-Century Analytic Philosophy* (Oxford: Blackwell).

Quine, W.V. (1960) *Word and Object* (Cambridge: MIT Press).

Quine, W.V. (1968) *Ontological Relativity* (New York: Columbia University Press).

Quine, W.V. (1974) *Roots of Reference* (La Salle: Open Court).

Quine, W.V. (1975) 'The Nature of Natural Knowledge'. In S. Guttenplan (ed.) 1975, *Mind and Language*. Oxford: Oxford University Press .

Quine, W.V. (1976) *Ways of Paradox*, revised and enlarged edition (Cambridge: Harvard University Press).

Quine, W.V. (1980 [1953]) *From a Logical Point of View* second edition revised (Cambridge: Harvard University Press).

Quine, W.V. (1981) *Theories and Things* (Cambridge: Harvard University Press).

Quine, W.V. (1992 [1990]) *Pursuit of Truth*, second edition (Cambridge: Harvard University Press).

Quine, W.V. (1995) *From Stimulus to Science* (Cambridge: Harvard University Press).

Quine, W.V. (1998) *The Philosophy of W.V. Quine*, second edition (La Salle: Open Court).

Quine, W.V. (2008) *Confessions of a Confirmed Extensionalist and Other Essays* (Cambridge: Harvard University Press).

Russell, B. (1951) 'Santayana's Philosophy'. In P.A. Schilpp (ed.) 1951, *The Philosophy of George Santayana*. La Salle: Open Court .

Waismann, F. (1979) *Wittgenstein and Vienna Circle* (Oxford: Blackwell).

Wittgenstein, L. (1990 [1922] *Tractatus Logico-Philosophicus*. Tr. C.K. Ogden (London: Routledge).

Wittgenstein, L. (1953) *Philosophical Investigations* (Oxford: Blackwell).

Wittgenstein, L. (1958) *The Blue and Brown Books* Oxford: Blackwell.

Wittgenstein, L. (1966) *Wittgenstein: Lectures and Conversations* (Berkeley: University of California Press).

Wittgenstein, L. (1977) *Remarks on Colour* (Oxford: Blackwell).

Wittgenstein, L. (1980) *Remarks on the Philosophy of Psychology,* volume 1 (Oxford: Blackwell).
Wittgenstein, L. (1993) *Philosophical Occasions 1912–1951* (Indianapolis: Hackett).
Wittgenstein, L. (1998) *Culture and Value* (Oxford: Blackwell).
Wittgenstein, L. (2005) *The Big Typescript: TS 213* (Oxford: Blackwell).

Appendix

Scan of Quine's Original 'Pre-Established Harmony'

PREESTABLISHED HARMONY

W. V. Quine

In my ill-organized way I have only now, a year late, come upon Gary Ebbs' brilliant review of my Pursuit of Truth.[1] I am flattered by his scholarly command of my writings and impressed with his penetration to crucial points. I am thankful, amid all this, for his evidently having missed one vital point in my evolving views; for other readers will have missed it a fortiori, and I am now alerted to make amends for my inadequate exposition. The point he seems to have missed is the preestablished intersubjective harmony of perceptual similarity standards, rooted in natural selection.

It bears on the learning of observation sentences, by the child in the home language and the field linguist in the alien language. According to Word and Object, to learn an observation sentence is to endow it with a "stimulus meaning" similar to those that it has for other speakers. A stimulus meaning was a set of sets of nerve endings, and I was uneasy even then about intersubjective similarity of such sets. I sketched the field linguist's actual procedure in terms purely of observation of behavior, then as now; the linguist was never meant to know about nerve endings and stimulus meaning. By 1990 I was applying the buzz word 'empathy' to this routine, but it was the same old behavioristic routine.

In 1986, meanwhile, it had tardily dawned on me that since the child and the field linguist equate observation sentences intersubjectively on the basis purely of observable behavior anyway, the privacy of stimulus meanings can be left inviolate, intersubjectively walled off.[3]

---

[1] Philosophical Review 103 (1994), pp. 535-541.

[2] Pursuit of Truth (Cambridge: Harvard, 1990), pp. 42?

[3] Ibid., p. 41.

2

If my word 'empathy' hinted of a mentalistic turn, I aggravated the suspicion by writing that the

> handing down of language is ... implemented by a continuing command, tacit at least, of the idiom 'x perceives that p' where 'p' stands for an observation sentence. Command of this mentalistic notion would seem therefore to be about as old as language. It is remarkable that the bifurcation between physicalistic and mentalistic talk is forewhadowed already at the level of observation sentences, as between 'It is raining' and 'Tom perceives that it is raining'. Man is indeed a forked animal.[4]

It is unwarranted, however, to read this as endorsing mentalism. More than once I have remarked on the serendipitous fruits of confusions.[5]

As for the propositional attitudes, of which I view 'perceives that' as the pioneer, I have long recognized and deplored my inability to get along without them.[6] In this matter I reached my present _modus vivendi_ in 1992.[7] As of now I cling to extensionalism, and thus to classical predicate logic as the logic of acceptable scientific discourse. I have reconciled the propositional attitudes _de dicto_ with extensionalism, via quotation and spelling, and reconciled myself to banishing propositional attitudes _de re_. These languish in the limbo of auxiliaries, along with the indicator words.

This leaves the mentalistic predicates of propositional attitude aboard, but no breach of extensionality. Such is my commitment to Davidson's anomalous monism. Scientific language thus inclusively conceived tolerates these predicates, albeit as epistemological danglers

---

[4] Ibid., pp. 61-62

[5] E.g. in _Roots of Reference_ (LaSalle, Ill.: Open Court, 1974), pp. 68, 125.

[6] Thus my "double standard" in _Word and Object_, pp. 216-221.

[7] Pursuit of Truth, 2d ed. (1992), pp. 70.

that neurology and physics can do without. Paul Churchland dreams of reducing them to neurology, and I hardly need say that I should be more than pleased.

Ebbs noted most of the foregoing points in his review. I have now to make my main one. I explained that my use of the word 'empathy' invokes nothing beyond what the mother and the field linguist were already up to according to Word and Object. There remains, however, in both the setting of Word and Object and that of Pursuit of Truth, a question of causal explanation.

Take the case of mother and child; the other case is parallel. Word and Object has the mother inducing in the child a stimulus meaning for the observation sentence 'Milk' similar to her own. Similar stimulus meanings cause similar responses to similar stimuli, so the child's subsequent use of the observation sentence agrees with the mother's. But this causal explanation appeals to intersubjective similarity of stimulus meanings.

In Pursuit of Truth that appeal is out of order, so the causal question recurs: why, after the mother has got the child to affirm 'Milk' once in an appropriate situation, does the child's usage continue to agree with the mother's? The answer lie can no longer in intersubjective similarity of stimulus meaning. It now lies rather in an intersubjective parallelism of subjective scales of perceptual similarity. If A and B jointly witness two events, and A's neural intakes on the two occasions are perceptually similar by A's standards, then B's intakes will ~~likewise~~ tend to be similar by B's.

Intersubjective harmony of similarity standards is needed not only in accounting one presentation of milk similar to another, but also in accounting one utterance of 'Milk' similar to another. Failing such harmony, the child's continuing heralding of milk would fall on uncomprehending ears.

4

The harmony of innate standards of perceptual similarity is accounted for by natural selection. Similarity is the basis of expectation, for we have an innate tendency to expect similar events to have sequels similar to each other. This is primitive induction. Accordingly a scale of perceptual similarity has survival value insofar as it is conducive to successful expectation, and hence to anticipation of predator, prey, and other threats and boons. Shared environment down the generations would make, then, for parallel similarity scales. Difference of environment would make eventually for difference in similarity scales between different peoples, but the lot of humanity around the world down the ages has been enough alike to make for parallelism of evolution in its main lines.

Darwin wrought the great revolution in metaphysics, out-distancing Copernicus; for he reduced final cause to efficient cause. Now we see his theory at work in epistemology, affording a naturalistic account of our seeming access to other minds.

Scan of Quine's Original 'Reply to Gary Ebbs'

RESPONSE TO GARY EBBS

W. V. Quine

In my ill-organized way I have only now, a year late, come upon
Gary Ebbs' brilliant review of my Pursuit of Truth.[1] I am flattered
by his scholarly command of my writings and impressed with his pene-
trati to crucial points. I am thankful, amid all this, for his evidently having missed certain vital points in my evolving views; for
other readers will have missed them a fortiori, and I am now alerted
to make amends for my inadequate exposition.

Ebbs sees the transition from Word and Object (1960) to Pursuit
of Truth (1990, 1992) as adulterating my naturalism with mentalism.
He infers this from my abandonment of the intersubjective matching of
stimulus meanings and my resort to empathy.

'Empathy' has a mentalistic ring, but the procedures that it denotes in Pursuit of Truth (p. 42) are just the observations of behavior
that I had ascribed to the field linguist in Word and Object (pp. 29-
30). The linguist was never meant to have any conception of stimulus
meanings or of the nerve endings of which they are composed. The notion of stimulus meaning does not even emerge until a later page (32).
Empathy as it figures thus anonymously in Word and Object and by name
in Pursuit of Truth is mentalistic only in the negative sense of being
manifested in behavior rather than defined physiologically. Appeal
to it is no breach of naturalism or physicalism, by my lights.

If my word 'empathy' hinted of a mentalistic turn, I aggravated
the suspicion by writing that the

handing down of language is ... implemented by a continuing command,
tacit at least, of the idiom 'x perceives that p' wherev'p' stands

---

[1] Philosophical Review 103 (1994), pp. 535-541.

2

for an observation sentence. Command of this mentalistic notion would seem therefore to be about as old as language. It is remarkable that the bifurcation between physicalistic and mentalistic talk is forewhadowed already at the level of observation sentences, as between 'It is raining' and 'Tom perceives that it is raining'. Man is indeed a forked animal. [Pursuit of Truth, pp. 61-62]

It is unwarranted, however, to read this as endorsing mentalism. More than once I have remarked on the serendipitous fruits of confusions.[2]

As for the propositional attitudes, of which I view 'perceives that' as the pioneer, I have long recognized and deplored my inability to get along without them.[3] In this matter I reached my present modus vivendi in 1992.[4] As of now I cling to extensionalism, and thus to classical predicate logic as the logic of acceptable scientific discourse. I have reconciled the propositional attitudes de dicto with extensionalism, via quotation and spelling, and reconciled myself to banishing propositional attitudes de re. These now languish in the limbo of auxiliaries, along with the indicator words.

This leaves the mentalistic predicates of propositional attitude aboard, but no breach of extensionality. Such is my commitment to Davidson's anomalous monism. Scientific language thus inclusively conceived tolerates these predicates, albeit as epistemological danglers that neurology and physics can do without. Paul Churchland dreams of reducing them to neurology, and I hardly need say that I should be more than pleased.

2 E. g. in Roots of Reference (LaSalle, Ill.: Open Court, 1974), pp. 68, 125.

3 Thus my "double standard," Word and Object, pp. 216-221.

4 Pursuit of Truth, 2d ed., pp. 65-72.

It seems in several passages that Ebbs takes my term 'science' too narrowly. We have no word with the breadth of <u>Wissenschaft</u>, but that is what I have in mind. History is as at home in my naturalism as physics and mathematics. So also, indeed, is translation. My conjecture of indeterminacy of translation is just that in the radical translation of theoretical material there may be incompatible alternative turnings and no fact of the matter: either will serve, but not both. This is a caveat regarding the notion of meaning, and not to be read as <u>traduttori traditori</u>.

Ebbs sees my privatizing of stimulus meanings, in passing from <u>Word and Object</u> to <u>Pursuit of Truth</u>, as more drastic than I do. *Stimulus meanings* still figure in <u>Pursuit</u>, *and* they never were meant to figure in the practice of translation or language teaching. Their role early and late was causal: they are the neural launching pads of observation sentences.

Intersubjective likeness of stimulus meanings served, in its day, to account causally for our continuing *Intersubjective* agreement in the affirmation of an observation sentence from occasion to occasion. The child's usage and the mother's, or the linguist's and the native's, do not drift apart along divergent paths of extrapolation. Take the case of mother and child; the other case is parallel. <u>Word and Object</u> has the mother inducing in the child a stimulus meaning for the observation sentence 'Milk' similar to her own. Similar stimulus meanings cause similar responses to similar stimuli, so the child's subsequent use of the observation sentence continues to agree with the mother's. The stimulus meaning is a set of sets of receptors, and the relative stability of the receptors would arrest drift. If the child's stimulus meaning of the observation sentence matched the mother's (or the linguist's the native's) on the first occasion, it will continue to.

Still, even at the time of <u>Word and Object</u>, I was uneasy about this

4

intersubjective matching of stimulus meanings. It called for an intersubjective homology or near-homology of nerve endings that I felt ought
to be irrelevant. Davidson, Dreben, and Føllesdal pressed the problem
at our little conference in 1986. At length it dawned on me how else
to account for continuing agreement over observation sentences.

It is due rather to a preestablished harmony of subjective scales
of perceptual similarity. If A and B jointly witness two events, and
A's neural intakes on the two occasions are perceptually similar by A's
standards, then B's intakes will likewise tend to be similar by B's.

An individual's standards of perceptual similarity can in principle be elicited esperimentally by the reinfornement and extinction
of responses.[5] They are largely unlearned, since learning depends on
them; but they change gradually with experience.

The harmony of innate standards of perceptual similarity is accounted for by natural selection. Similarity is the basis of expectation,
for we have an innate tendency to expect similar events to have sequels
similar to each other. This is primitive induction. Accordingly a scale
of perceptual similarity has survival value insofar as it is conducive to
successful expectation, and hence to anticipation of predator, prey, and
other threats and boons. Shared environment down the generations would
make, then, for parallel similarity scales. Difference of environment
would make eventually for difference in similarity scales between different peoples, but the lot of humanity around the world down the ages
has been enough alike to make for parallelism of evolution in its main
lines.

Darwin wrought the great revolution in metaphysics, out-distancing
Copernicus; for he reduced final cause to efficient cause. Now we see
his theory at work in epistemology, affording a naturalistic account of
our seeming access to other minds.

---

[5] See Roots of Reference, pp. 16-18. In the present paper I am indebted
to Burton Dreben for helpful discussion.

Index

219

Lightning Source UK Ltd.
Milton Keynes UK
UKOW06n1202210616

276769UK00005B/43/P